Guidance for Professional Development in Drinking Water and Wastewater Industry

Guidance for Professional Development in Drinking Water and Wastewater Industry

Archis Ambulkar

IWA PUBLISHING

Published by **IWA Publishing**
Alliance House
12 Caxton Street
London SW1H 0QS, UK
Telephone: +44 (0)20 7654 5500
Fax: +44 (0)20 7654 5555
Email: publications@iwap.co.uk
Web: www.iwapublishing.com

First published 2015

British Library Cataloguing in Publication Data
A CIP catalogue record for this book is available from the British Library

ISBN: 9781780406961 (Paperback)
ISBN: 9781780406978 (eBook)

Contents

List of Figures

About the Author

Mr Archis Ambulkar is an internationally recognized environmental engineering expert with work focused on water and wastewater systems. His contributions to the field have been recognized by several newspapers and technical publications. He has been involved as an expert in various capacities for organizations such as the United Nations (UN) World Ocean Assessment Program, the Food and Agriculture Organization of the United Nations (UN FAO) Global Soil Partnership Initiative, the Intergovernmental Panel on Climate Change (IPCC), the Water Environment Federation (WEF), the American Water Works Association (AWWA) and the International Union for Conservation of Nature (IUCN).

Mr Ambulkar has obtained a Master of Science degree in Environmental Engineering from Bucknell University, Pennsylvania, USA, a Bachelor of Technology degree in Chemical Engineering from Nagpur University, India and has more than a decade of experience in the research and consulting fields. He has co-authored numerous publications including research papers, industry standards, factsheets, manuals of practices (MOPs), technical articles and case studies related to water, wastewater and waste management, published in countries including the USA, United Kingdom, Canada, China and India.

Mr Ambulkar has extensively served on the editorial boards of various scientific journals and technical magazines published in the USA, United Kingdom, Canada and India, including organizations such as the International Water Association (IWA), the American Society for Testing and Materials (ASTM) International and the Pennsylvania Water Environment Association (PWEA). His editorial involvements include various roles as Editor, Associate Editor, Editorial Board Member and Technical Advisory Board Member. As an Expert Reviewer Mr Ambulkar has made contributions to the Intergovernmental Panel on Climate Change's (IPCC) Reports, reviewed more than 150 research manuscripts published by Elsevier Publishing, IWA Publishing and WEF Publishing. He has also served as a peer-reviewer for technical manuals published by WEF.

In addition to being an expert member with several technical Committees with AWWA, WEF and ASTM International, Mr Ambulkar has served as a co-chairman for the Water Environment Federation Technical Exhibition and Conference's (WEFTEC) Annual Career Fair Events and as a jury member for the Water Industry Hall of Fame Awards selection process for the American Water Works Association. He has been appointed as a technical committee member for several conferences organized by Asia-Pacific Chemical, Biological and Environmental Engineering Society (APCBEES) in UAE, Hong Kong and New Zealand.

Mr Ambulkar was elected as a fellow of the Linnean Society of London (United Kingdom) in 2013 and awarded 'Man of the Year 2013' by the International Biographical Centre of Cambridge, England in recognition of services to the environmental engineering field. He is also recognized and listed in Marquis Publication's 'Who's Who in the World', 'Who's Who in America' and 'Who's Who in the East'.

Preface

With growing environmental awareness and rising concerns for sanitation, hygiene, drinking water quality, water conservation and public health, governments, private institutes as well as international organizations are undertaking major drinking water and wastewater infrastructure projects. Correspondingly, the demand for highly qualified professionals to perform research work, develop new technologies as well as plan, design and execute these projects has increased globally.

As several technical books are available in the market related to planning, designing, operations and management of water and wastewater systems, the present book focuses on professional development aspects for aspirants starting from student phase, through junior level position all the way up to a well-established expert status. This book will provide guidance for each step of the professional journey and is supported by a comprehensive technical database necessary to understand the profession and jumpstart a career in the water or wastewater industry. This publication will serve as a bridging document between introductory books and highly specialized technical books or publications related to drinking water or wastewater systems.

It is believed that the book will be an excellent read for prospective students and professionals. Also, that it can be effectively used by companies as a training tool, by university or college staff members for educational purposes as well as by well-established experts to guide newcomers or junior staff working in the drinking water and wastewater field.

Acknowledgement

This book would not have been possible without sustained support of my family. I am indebted to them for supporting throughout my career in all aspects. I would like to dedicate present publication to my family members, educational institutes, mentors and colleagues who had a great influence on my personality and on my path to become an established environmental professional. I am also thankful to various organizations, institutes and authors for granting permission to refer their publications for developing this book.

Chapter 1

Introduction

Rising world population, growing demands and resulting scarcity of land, energy and water are well known facts of today. In addition to several ongoing efforts for exploring alternative energy resources, efficient land use methods and improving air quality, water conservation and its reuse have gained significant attention. Also, improper sanitation, unauthorized wastewater discharges, consumption of non-potable water and subsequent diseases or deaths are major areas of concern worldwide and especially in poor countries.

As per the World Health Organization (WHO) report 'Safer Water, Better Health – Costs, benefits and sustainability of interventions to protect and promote health' (WHO, 2008), '*an important share of the total burden of disease worldwide – around 10% – could be prevented by improvements related to drinking-water, sanitation, hygiene and water resource management.*' Millions of deaths and infections are preventable in this manner. Figure 1.1, from the report, summarizes various diseases contributing to the water-, sanitation- and hygiene-related disease burden.

1.1 GLOBAL DRINKING WATER AND SANITATION SCENARIO

The 'Progress on Drinking Water and Sanitation 2012 Update' report by The United Nations Children's Fund (UNICEF) Organization and WHO, provides global scenario for drinking water and sanitary systems (UNICEF/WHO, 2012). The report suggests that over 780 million people are still without access to improved drinking water sources and 2.5 billion lack improved sanitation. Figure 1.2 from the report depicts that sub-Saharan Africa countries have the lowest drinking water coverage of any region in the world. Also, as far as sanitation systems are concerned, there are many countries (mostly sub-Saharan African and Southern Asian countries) with sanitation coverage below 50 per cent (Figure 1.3).

Water is an essential element for the existence of life. Lack of sufficient water sources accompanied by urbanization, deforestation and desertification activities have resulted in major water challenges for societies (Laureano, 2010; Smakhtin *et al.* 2008; Tollan, 2002). Climate changes, unpredictable weather conditions and changing precipitation patterns place demands on conservation and efficient use of available water resources (Arnbjerg-Nielsen *et al.* 2013; Nuorteva *et al.* 2010; Olsson, 2007). In short, smarter use of water and effective wastewater treatment with recycling options have become a priority for the well-being of civilization, as never before.

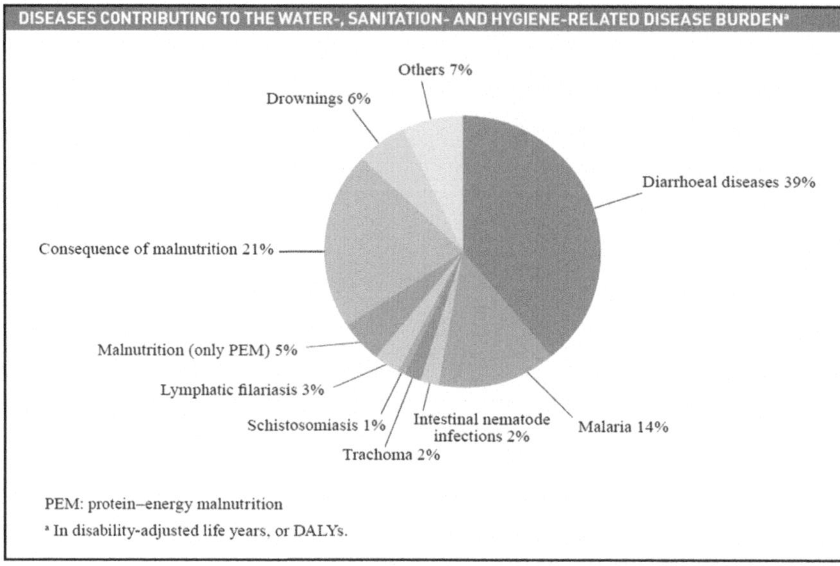

Figure 1.1 Diseases contributing to the water-, sanitation- and hygiene-related disease burden. (Reproduced with the permission of the publisher, from WHO, 2008.)

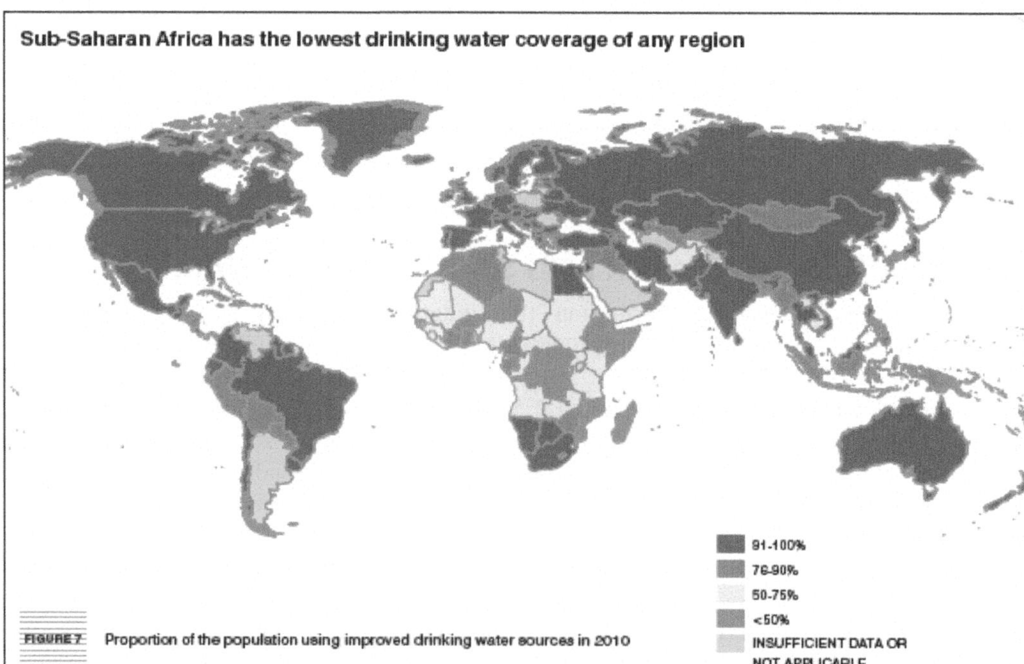

Figure 1.2 Population using improved drinking water sources in 2010. (Reproduced with the permission of the publisher, from UNICEF/WHO, 2012.)

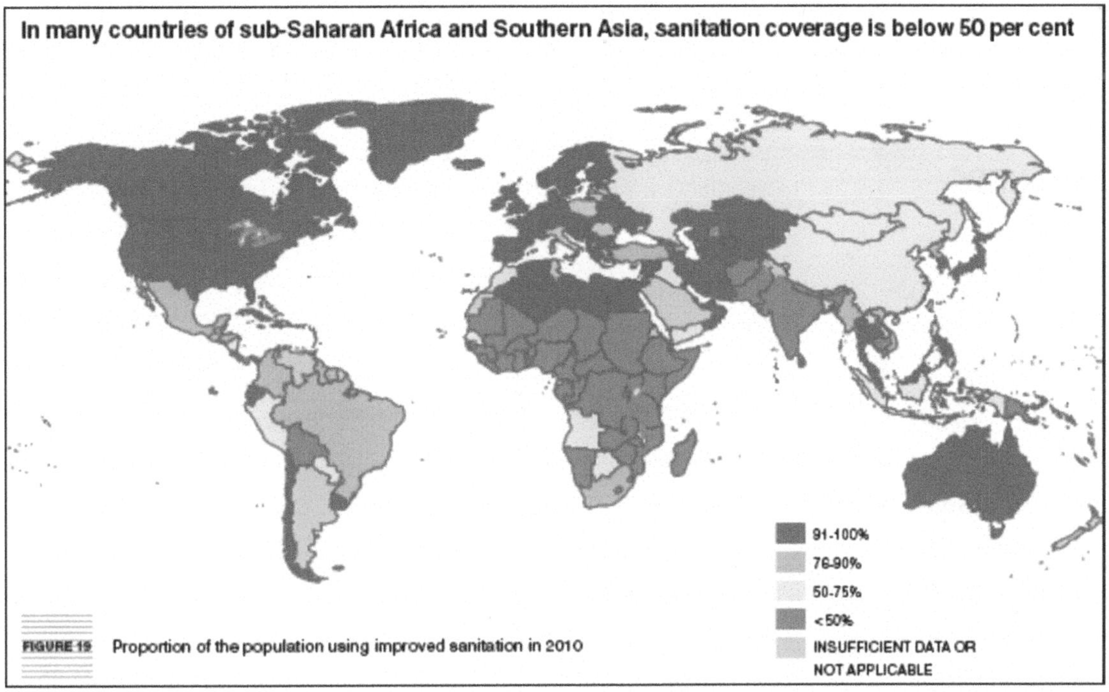

Figure 1.3 Population using improved sanitation in 2010. (Reproduced with the permission of the publisher, from UNICEF/WHO, 2012.)

With growing concerns at a global level, several international organizations are promoting improvements to drinking water and sanitary systems to achieve healthy and hygienic conditions in different parts of the world. Also, national governments and regional municipal associations are undertaking water and wastewater related infrastructure projects and allocating funds for their implementation.

To develop newer technologies, advance scientific knowledge and manage infrastructure improvement projects, well-educated and experienced drinking water and wastewater professionals are needed. Hence, education in the environmental field followed by a career in this profession has become an important topic of the present times.

1.2 DRINKING WATER AND WASTEWATER PROFESSION

Amongst various technical professions such as chemical, mechanical, civil, electrical, industrial, nuclear, mining, metallurgical, and so on, the environmental field enjoys a 'special visionary status' and deals with preservation and improvement of the natural environment with specific focus on water, air and land. Figure 1.4, is a typical water cycle showing its transition from one form to other in nature (United States Geological Survey, USGS).

As far as professional scope is concerned, drinking water systems typically include water sources (such as rivers, lakes, reservoirs, groundwater aquifers, etc.), intake structures, treatment systems, storage facilities and community distribution networks (Alaska DEC, 2013; U.S. EPA, 2004; U.S. Fire Administration, 2008). Whereas, wastewater systems commonly include domestic and industrial wastewater collection and conveyance systems, onsite, centralized or decentralized wastewater treatment plants, sludge management systems and methods for discharging treated effluent water back to the above-ground sources (such as streams, rivers, lakes, oceans, etc.) or to water aquifers in the ground via underground discharge (Libabher & Orozco-Jaramillo, 2012; U.S. EPA., 2000b,e, 2002a,b, 2003a, 2005a, 2006b, 2013; van Haandel *et al.* 2012).

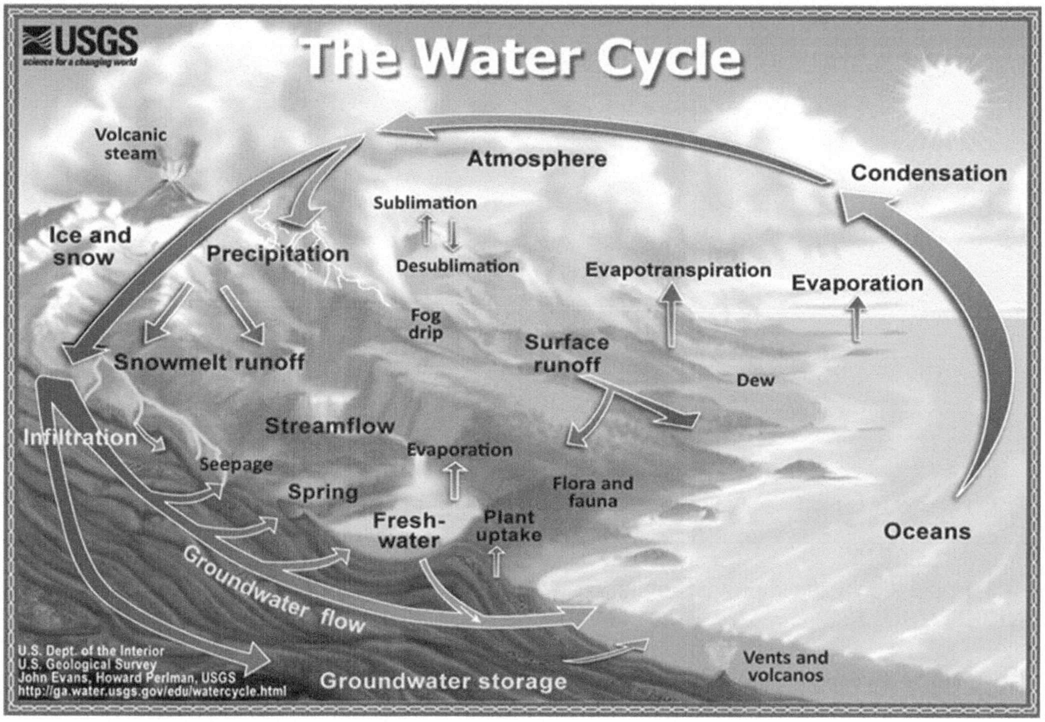

Figure 1.4 Water Cycle Diagram. (Reproduced with the permission of the publisher, from U.S. Geological Survey.)

In general, drinking water and wastewater professional activities involve scientific research, technological advancements, academic teaching, laboratory work, policy making, planning processes, development of industry standards and execution of field projects involving surveying, designing, permitting, construction, operations, maintenance and management. Depending upon the regions, countries, regulatory requirements, funds availability, community needs, priorities, system sizes, generation quantities, demands and several other factors, professional scope and involvement can vary significantly.

Due to the growing environmental concerns and implementation of stringent limits for water quality improvements and pollution reduction, the need to research sophisticated technologies, construct newer systems, upgrade existing infrastructure is rising internationally. The ability to manage and execute such projects comes with a financial burden on organizations, institutions, societies and the public in general. Proper implementation of tasks requires specialized experts with appropriate education, knowledge and experience in the field.

Availability of qualified professionals to satisfy growing needs is an important aspect. Hence, it is imperative that aspirants get first-hand knowledge and understanding of the drinking water and wastewater profession while selecting their career path. Also, beforehand guidance on key professional aspects can help recent graduates, entry-level and mid-level professionals pursue work efficiently and minimize the learning time needed to establish themselves in their career.

Often, readers are expected to have in-depth subject knowledge in order to grasp technical and scientific publications by experts as well as profession-related documents. To inculcate interest in the environmental field and encourage prospective students and professionals, a bridging document is required to introduce them to the basics of this profession, familiarize them with job-related aspects and provide an overview of challenges and opportunities within the industry.

With several books by professionals related to design, operations, management and research available in the market, a publication providing an overview of the profession and advice for individual growth within the industry was thought to be a useful contribution towards the field. Hence, the present book is a humble attempt to offer the guidance necessary for professional development in the drinking water and wastewater sector from the prospective student stage to the established practitioner's position.

1.3 PUBLICATION OBJECTIVES

Emphasizing drinking water and wastewater, this book realizes the practitioner's journey on both professional level and the technological front. As each individual's journey is different and unique in itself, the present publication outlines common career phases involved and helps realize the purpose, motivation, responsibilities and milestones of the various stages. Finally, the book helps individuals appreciate this occupation as a noble, enjoyable and fulfilling experience. The book is a valuable reference for university students, entry-level and mid-level professionals. It can also be used by established seniors, mentors and university professors as a training tool or specific sections can be used to guide newcomers or junior staff members.

1.4 SCOPE AND GOALS

Delineating the scope and goals was a challenging task during the development process, mainly because of the availability of a wide range of books, manuals, standards, and so on, on drinking water and wastewater subjects. Considering the purpose of this publication, the scope mainly includes 'individual professional growth' from student to established professional level, the 'basic technical knowledge' necessary for understanding the profession, 'pursuing exceptional achievements' and finally an extensive database of key resources for quick development within the field.

The book also offers recommendations, special remarks and 'do's' and 'don'ts' of the profession. This narration will help readers realize a step-by-step professional development process in the industry and at the same time receive key inputs to minimize or avoid common mistakes related to the drinking water or wastewater occupations. Knowledge from this publication will be beneficial to accelarate individual's 'learning process' aimed at establishment in the field. With the combination of subjects handled, the present book will surely boost professional's growth prospects.

1.5 CHAPTER CONTENTS

The book's contents are broadly divided into chapters 'Educational Options', 'Career Avenues', 'Professional Development Overview', 'Knowing Drinking Water and Wastewater Systems', 'Understanding Drinking Water and Wastewater Projects', 'Realizing Exceptional Achievements' and finally 'Key Resource Databases for the Profession' as Appendices.

- Prior to initiating discussions for various professional phases, educational options and common career avenues in the drinking water and wastewater industry are discussed. Basic educational options include short-term courses, diploma, certificate, associate degree, bachelor's degree, master's degree, doctorate degree, postdoctoral research and continued educational options. Job options commonly include academic faculty member, researcher, consultant, engineer, designer, construction management staff, system operator, laboratory staff, permitting authority staff, manufacturing staff, utility company employee, industry staff or sales and marketing staff.

- The chapter covering professional development and growth from novice learner to an established practitioner is divided into stages such as 'Student Phase', 'Internship Phase', 'Entry-Level Professional', 'Mid-Level Professional' and 'Established Practitioner'. Sections on each phase include the learning processes involved, knowledge base development and special suggestions for realizing one's potential and becoming effective with the respective roles. Since career phases may vary on an individual basis and depend upon type of job profession pursued, general terminologies have been used for discussions.
    - **'Student Phase'** is marked by enthusiasm, energy and ideas, however with lack of knowledge and experience. By orienting strengths in the right direction, towards correct information and with appropriate motivation, high quality graduates can be developed. This chapter discusses aspects such as pursuing educational programs, curriculum and subjects, funding opportunities, interactions with faculty members and developing the knowledge base necessary for a drinking water or wastewater career.
    - **'Internship Phase'** is an intermediate step towards becoming a professional. This stage provides an appropriate launch pad with a 'comfort factor', as it helps to understand professional work environment while still pursuing a college degree. The chapter discusses internship roles with a focus on learning the basics of the business, developing communication skills, getting a glimpse of work experience and preparing for real-job as an entry-level professional.
    - **'Entry-Level Professional Phase'** can be one of the most nervous phases in the professional career. Although riding high on a recent graduate degree, the transition from a well-nurtured educational phase to a fully-independent work phase can sometimes be a struggle and somewhat cumbersome. This stage marks the beginning of a career, in which education gets accompanied by real work experience. The chapter contemplates topics like job search, application and interview process, starting an occupation, interdepartmental coordination, learning job-related documents, vendor interaction, managing deadlines, and so on.
    - **'Mid-Level Professional Phase'** is the next important stage in career growth. With considerable work experience on the resume, mid-level individuals serve two-pronged duties: on one side providing guidance to the junior staff and team members and on other side managing projects and giving key inputs to the top management for achieving company goals. This chapter discusses multi-tasking, project management, learning industry standards, professional designations, and involvement with key national and international organizations for growing eminent presence in the field.
    - **'Established Practitioner'** is an individual with substantial expertise and authority in the drinking water or wastewater profession who can provide vision and direction to the company, staff members and community as a whole. In general, these practitioners can be at various hierarchical levels in the company starting from established researcher, designer or engineer, all the way to company owner or executive member. The chapter discusses types of established professionals, significant transitions in a career, the practitioner's unique journey, mentoring and guidance, and so on.
- 'Drinking Water Systems' section marks the beginning of chapters on technical aspects. It discusses water treatment and distribution systems with a focus on specific aspects, such as water sources, intake systems, treatment processes, followed by the distribution network involving pipes, valves, hydrants, tanks and other components.
- 'Wastewater Systems' section discusses different wastewater collection and conveyance systems such as gravity sewers, pumping stations, vacuum sewers, pressure sewers, along with onsite, clustered or

centralized wastewater treatment processes involving septic tanks, screening, grit removal, chemical treatment, biological reactors, clarifiers, sludge digestion and other processes practiced in the industry.

- 'Understanding Drinking Water and Wastewater Projects' chapter deals with infrastructure projects with focus on 'Clients Overview', 'Projects Overview', 'Planning Phase', 'Designing Phase', 'Bidding and Construction' and 'System Operations and Management'.
- Upon reviewing the professional development and technical background for drinking water and wastewater systems, the last chapter 'Realizing Exceptional Achievements' discusses the 'special efforts' that will help professionals stand out distinctly amongst other peers and potentially become an inspiration and role model for the next generation within the industry. Several facets of special achievements such as 'Realizing Exceptional Goals', 'Making Profession a Passion', 'Creativity and Reasonable Risks', 'Achievements with Ethics', 'Struggle Again Odds' are discussed here.

Overall, the book aims to encourage an appreciation of the importance of the drinking water and wastewater industry, to recognize a career within it as a distinguished profession and to motivate individuals to pursue their passion with hard work in order that professional goals are achieved.

For the reader's convenience and ease of navigation through the text, Figure 1.5 provides the sequence of various chapters included in the book.

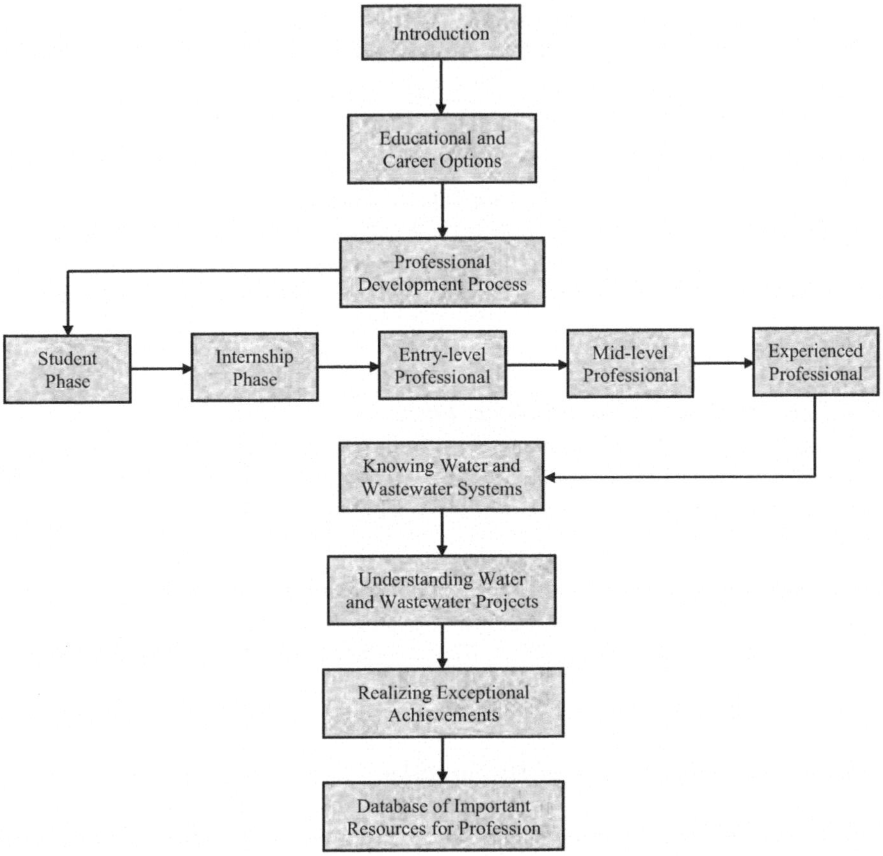

Figure 1.5 Overview of book chapters.

1.6 PROFESSION'S KEY RESOURCE DATABASES

In order to further enhance working knowledge of the drinking water and wastewater topics, a combination of key databases have been included as appendices in the book. These resources should prove beneficial for professionals and are summarized below for the reader's convenience.

• ***Appendix A – Glossary: Common Water and Wastewater Terminologies***
Appendix A includes commonly used water and wastewater terminologies and their meanings for reference. A first-hand knowledge of terms will come in handy in understanding projects, reading reports, studying technical topics and for professional conversations.

• ***Appendix B – Glossary: Construction Management Terms***
Appendix B includes common terminologies associated with construction management projects, their meaning and will be beneficial in reviewing and executing drinking water and wastewater construction projects.

• ***Appendix C – Glossary: Laboratory Operations Terminologies***
Drinking water or wastewater laboratory staff or professionals working in this area should be aware of the common terminologies associated with laboratory operations. Appendix C includes a list of common terms used and their meaning.

• ***Appendix D – Common Conversion Factors and Basic Formulae***
Unit conversions are important for various professional tasks such as calculations, designs, interpreting results, preparing drawings, and so on. Appendix D includes common conversion factors along with basic formulae in the water and wastewater field, such as area, volume, flow, pumping and power requirements for reference.

• ***Appendix E – Common Drinking Water System Formulae***
Quite often professionals need to use equations and formulae to design water treatment systems. Appendix E includes common drinking water treatment system formulae for processes such as water well pumping, coagulation, sedimentation, filtration, chlorination and chemical feed systems.

• ***Appendix F – Common Wastewater Treatment System Formulae***
Appendix F includes common wastewater treatment system formulae for processes such as primary treatment, clarification, activated sludge systems, tricking filters, sludge digestion, sludge thickening and dewatering.

• ***Appendix G – Common Laboratory Analysis Formulae***
Laboratory sampling and analysis are an important part of both drinking water and wastewater systems. Appendix G includes common laboratory analysis formulae for parameters such as biochemical oxygen demand, alkalinity, total solids, volatile solids, fixed solids, sludge volume index, bacterial colonies, and so on.

- ***Appendix H – Listing: Water Treatment System Components***

Water treatment systems typically include a series of unit operations and processes for producing high quality drinking water. These systems may include an influent pumping station, sedimentation, flocculation, adsorption, ion exchange, disinfection, and so on. Appendix H lists various basic components involved with these process units to further enhance the reader's practical knowledge about water treatment systems.

- ***Appendix I – Listing: Wastewater Treatment System Components***

Wastewater treatment processes involve several unit operations and processes in order to generate effluent water that can be safely discharged to surface or groundwater sources. These treatment systems may include an influent pump station, screening, grinding, grit removal, biological treatment, disinfection, sludge digestion, and so on, processes. Appendix I lists key components of common processes involved with the wastewater treatment systems.

- ***Appendix J – Listing: Chemicals Used for Treatment Processes***

Chemicals are used for various purposes during water or wastewater treatment operations such as coagulation, flocculation, precipitation, disinfection, oxidation, odor removal, pH adjustment, control taste, inhibiting biological growth, fluoridation, and so on. A basic understanding of these common chemicals will help readers to add to their working knowledge. A partial list of common chemicals used for drinking water and wastewater systems is included in Appendix J.

- ***Appendix K – Listing: Laboratory Equipment and Instruments***

Laboratory operations involve various equipment and instruments along with glassware, plastic-ware, utensils, sampling and analysis apparatus, appliances, furniture, safety gears, and so on. Appendix K lists a common inventory for the drinking water and wastewater laboratory operations.

- ***Appendix L – Listing: Common Construction Heavy Equipment***

Drinking water and wastewater infrastructure projects frequently use heavy equipment during construction such as dump truck, forklift, grader, compactor, bulldozer, concrete mixers, crane, excavator, loader, backhoe, and so on. A basic description of these equipment and their purpose is included in Appendix L.

- ***Appendix M – Listing: Water and Wastewater Profession's Key Gadgets and Tools***

Drinking water and wastewater professionals are involved in a wide range of routine tasks as a part of the profession that require specific key gadgets and tools. Appendix M lists such key resources and identifies common tasks to be performed for various project related activities.

- ***Appendix N – Water Project Cost Estimation Sheets***

Developing project costs is one of the common tasks encountered by drinking water professionals. New or upgrade projects related to treatment plants or distribution systems involve several aspects such as equipment purchasing and installation, earthworks, design services, construction management, mobilization and demobilization, bonds, insurance, contingencies, and so on. A simplified worksheet is included in Appendix N to assist with cost estimation for water projects.

- ***Appendix O – Wastewater Project Cost Estimation Sheets***

Project costs development is routinely exercised by wastewater professionals. New or upgrade projects related to collection and conveyance systems or treatment plants involve several aspects such as equipment purchasing and installation, earthworks, design services, construction management, mobilization and demobilization, bonds, insurance, contingencies, and so on. A simplified worksheet is included in Appendix O to assist with cost estimation for wastewater projects.

- ***Appendix P – Laboratory Project Cost Estimation Sheet***

Construction of new laboratory or upgrades to existing ones involves development of project cost estimates. These costs may include several factors such as laboratory building construction, lab furniture and administrative facilities, installation of laboratory equipment and instruments, safety equipment, chemicals, design services, construction management, earthworks, bonds, insurances, contingencies, and so on. A simplified worksheet is included in Appendix P to assist with the cost estimation for laboratory projects.

Chapter 2

Educational options in the drinking water and wastewater field

Chapter 2 discusses various educational options available in the drinking water and wastewater industry. Education is the basic foundation for any profession and needs to be taken very seriously. Subject understanding gained during studies goes a long way in a career and becomes the core knowledge base for the professional journey. Since educational structures can vary significantly from region to region and country to country, the purpose of the present chapter is to introduce readers to the various common options available for pursuing studies in the drinking water and wastewater areas (see Figure 2.1).

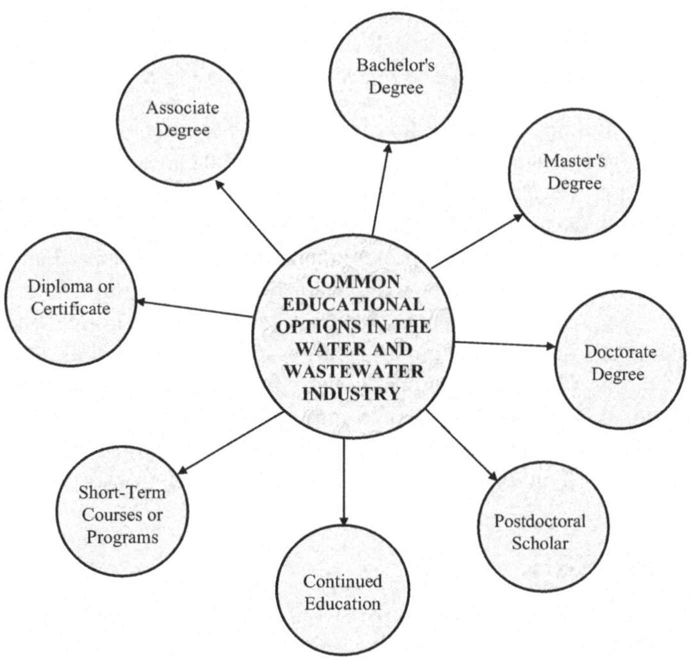

Figure 2.1 Common educational options in the drinking water and wastewater industry.

Now-a-days, a wide range of educational and learning avenues are available globally for gaining primary knowledge in the environmental field. These resources may include schooling, distance learning, online education, internet resources, organization websites, specialized software, libraries, community events, awareness programs, training sessions, television shows and many more. With a great deal of information available during the early educational phase, aspirants should make effective use of these resources for pursuing environmental studies. Once schooling is completed and the higher education stream is finalized (in this case the environmental, and specifically, drinking water or wastewater field), students can then select appropriate courses and degrees they want to pursue.

Diploma, certificate, associate's, bachelor's, master's and doctorate degrees are some of the common educational programs offered in the science and engineering field (NSB, 2014; The Sloan Consortium, 2005). The following narration provides a brief overview of these educational options with respect to drinking water and the wastewater profession.

2.1 SHORT-TERM COURSES OR EDUCATION PROGRAMS

Short-term introductory courses or specialized education programs related to the environmental field and particularly drinking water or wastewater systems are important resources for developing adequate understanding of the subjects in a short time and with minimum commitments. Upon completion of assignments, students can quickly get an idea of what the field has to offer in terms of profession and career. If sufficient interest gets developed in drinking water or wastewater topics, an individual can then proceed with advanced studies and an educational degree in the area offered by universities, colleges or other institutions. Prior to pursuing these short-term courses, one should make sure that the selected organizations and courses offered are legitimate and authentic. These short-term courses or training can be the first and easiest step towards gaining useful knowledge in the environmental field.

2.2 DIPLOMA OR CERTIFICATE

Although there are several avenues available for pursuing higher studies, factors such as career goals, lifestyles, family commitments, educational fees, personality, financial and other constraints can have an impact on a student's decision for selecting the desired degree and institution (CRA-W, 2000). Diploma or Certificate options provide the necessary study platform, qualifications as well as basic and applied knowledge in the area for students or working professionals that can eventually help with pursuing higher degrees or obtain jobs in the drinking water or wastewater field. Many positions such as field technicians, laboratory assistant, and so on, may require professionals to obtain diploma or certificates in the respective areas. For working professionals interested in pursuing advance knowledge in this field, part-time education, correspondence courses, online studies or evening classes can be useful options. In many cases, instead of joining universities for degree coursework, aspirants may pursue diploma or certificates from institutions providing these facilities. Online learning avenues involve courses in which most or all of the content is delivered online and typically have no face-to-face meetings (The Sloan Consortium, 2005).

2.3 ASSOCIATE DEGREE

Similar to the diploma or certificate, an associate degree in the water or wastewater field can be a valuable option for students. Although degree requirements can vary worldwide, typically Associate degrees are 2-year programs offered at colleges. In certain cases students can continue their education further at colleges or universities and subsequently earn higher degrees (NSB, 2014; The Sloan Consortium, 2005).

This degree typically involves a detailed curriculum with requirements for successful completion of several environmental subjects to obtain the degree. An associate degree can be quite handy for individuals who want to enter the water or wastewater field but who cannot commit time for a Bachelor's or higher degree. There are many semi-skilled to skilled jobs requiring subject knowledge but not necessarily higher degrees. Graduates with an associate degree can fit in such positions. Students can always plan to pursue higher education in the future. Associate degrees can also be available in the form of classroom or online courses depending upon institutions and universities (The Sloan Consortium, 2005). It is critical to ensure that these degrees are recognized in the field and valid for obtaining jobs, prior to selecting any specific education institute and course curriculum.

2.4 BACHELOR'S DEGREE

A bachelor's degree is one of the most prevalent degrees in science and engineering (NSB, 2014). For students with higher educational goals and a desire to obtain advanced knowledge in the environmental field, a bachelor's degree is the right option. This degree involves rigorous course work supplemented by projects, seminars, quizzes, and so on. Students get exposure to various environmental subjects and particularly in the drinking water and wastewater areas. Being the foundation for pursuing jobs in the skilled worker category, a bachelor's degree is a great platform for the individual to learn fundamentals in the field and get detailed hands-on knowledge of basic and applied sciences and technologies from the coursework. During college studies, writing assignments and/or oral presentations require students to identify specific problems/issues and devise ways to resolve them with professors typically providing guidelines and format (AACU, 2007). Projects and seminars help students gain the ability to present and defend one's work, improve communication skills and develop the confidence needed for entering the real job environment.

2.5 MASTER'S DEGREE

Master's is the next higher degree and a step towards a doctorate degree (NSB, 2014). With desired subject knowledge gained from a bachelor's degree, this educational avenue becomes more specific in nature. Master's degree candidates are required to take specialized courses in water and wastewater areas and are also expected to complete a dissertation, thesis or project on a particular topic. With this requirement, a master's degree is usually detailed and focused on selected topics. Certain universities offer a master's degree option that does not require a thesis and in such case students are expected to take more course work to compensate for it (CRA-W, 2000). A master's degree by research is a better option for students planning for a doctorate degree, whereas students preparing to work in the industry may pursue a master's degree without the thesis option (CRA-W, 2000). The degrees in environmental engineering or sciences can be referred to as Master of Science (M.S. or M.Sc.), Master of Technology (M.Tech.), Master of Engineering (M.E.), Master of Philosophy (M.Phil.) and similar other titles that can vary in different regions and countries. Students successfully completing this degree typically get hired at key positions in the companies and are expected to handle skilled and challenging tasks in the drinking water or wastewater projects. Master's degree education is respected well in the industry. It gives the degree holder more credibility amongst peers (CRA-W, 2000).

2.6 DOCTORATE DEGREE

An advanced degree is viewed as a distinguishing criterion for separating job seekers in the pool of applicants and provides research and writing experiences essential for launching a career onto the fast track. The degree generally results in earning higher salaries from the very beginning of career (CRA-W, 2000). A doctor of philosophy degree is highly respected in the industry and commonly referred to as Ph.D. or D.Phil. Those with doctorates are specialized and highly skilled professionals who assist with advancing knowledge in the field by making contributions towards research and innovations (NSB, 2014). To pursue a Ph.D., students are expected to make high level of original contributions in the specific subject and publish research papers. Such candidates work very closely with their guide and professors for the development of a doctoral thesis (CRA-W, 2000). Very detailed in nature, Ph.D. studies involve spending several years learning and researching selected topics and defending thesis work successfully in front of the faculty members or department to obtain the degree. Job openings for Ph.D. professionals can sometimes be very specific in nature. These members often get involved in tasks such as advancing technologies, research based projects, R&D laboratories, pilot-scale studies, as faculty members at institutes and many other top corporate positions. A doctoral degree prepares a new generation of professional researchers in academia, industry, and government, as well as a highly skilled workforce for other sectors of the economy (NSB, 2014).

2.7 POSTDOCTORAL SCHOLARS

Upon completion of a doctoral degree, postdoctoral fellowships provide an intermediate platform for professionals to gain training and pursue a career in their area of interest. According to the National Postdoctoral Association (NPA), USA 'A postdoctoral scholar ("postdoc") is an individual holding a doctoral degree who is engaged in a temporary period of mentored research and/or scholarly training for the purpose of acquiring the professional skills needed to pursue a career path of his or her choosing.'

As discussed in the 'Finding the Perfect Post-Doc Position for You' article (Manahan, NPA), not all post-PhD careers require a postdoc. While pursuing a postdoc position, selecting specific research field is important. Aspirants may plan to pursue postdoc in the same field as that of their thesis work or can change their area of interest. It is a good idea to talk with advisors and discuss shortlisted institutes. An institution's background, research group history, financial aspects (such as available funds, salaries and benefits), work environment, visa sponsorships, and so on, are some of the important factors that candidates should review prior to pursuing specific postdoc positions (Gamberi & Scholes, NPA).

Postdoc positions are typically supervised by at least one senior scholar who actively promotes a postdoc's professional development. Postdocs working in the water and wastewater area make key contributions towards advancing research work in the field and publish their findings or results in scientific and research-based environmental journals.

2.8 CONTINUED EDUCATION

Even after completing their education and joining the workforce, individuals are expected to continue their education in the form of seminars, advanced courses, obtaining and maintaining certifications or licenses, training and other professional platforms to keep updated with the current knowledge in the field and learn about recent advances made in the drinking water and wastewater areas. Companies and institutions encourage professionals for such continued educational programs and many times sponsor associated educational and traveling fees.

2.9 PROGRAM REQUIREMENTS AND SPECIAL SKILLS

Although several educational options are discussed above, starting from basic courses or certifications all the way to doctorate degrees and post-doctoral research work, the specific program requirements related to the number of courses, educational credits, duration of studies, seminars, projects, theses, and so on can vary significantly for institutions worldwide. Prospective students are encouraged to review specific obligations thoroughly with the institutions before pursuing specific educational options.

It should be realized that apart from the coursework pursued and knowledge obtained, students also gain a variety of other skills and qualities. These may include the ability to function in different environments and roles, public speaking experience, computer literacy, teaching skills, interview skills, managing studies and projects, implementing complex research projects, organizational and analytical skills, understanding statistics, critical evaluation, a problem-solving approach, the ability to acknowledge differing views of reality, and many others (AGU, 2001).

2.10 TRENDS IN SCIENCE AND ENGINEERING

As discussed by the Association of American College and Universities in its report 'College Learning for the New Global Century', 2007, the world is being dramatically reshaped by scientific and technological innovations, global interdependence, cross-cultural encounters, and changes in the balance of economic and political power (AACU, 2007). The National Science Board's 'Science and Engineering Indicators 2014' highlights some of the major developments in international and U.S. science and engineering field (NSB, 2014). As per the report, globalization of higher education continues to expand. Universities in several countries have increased their enrollment of foreign science and engineering students. Higher education is undergoing rapid transformation. The growth of distance and online education through online courses and similar innovations is expanding access to knowledge. In both the United States and the rest of the world, the science and engineering (S&E) workforce has experienced strong growth over time. Innovation based on S&E research and development (R&D) is globally recognized as an important vehicle for a nation's economic growth and competitive advantage, hence growing numbers of workers worldwide are getting engaged in research (NSB, 2014).

Chapter 3

Career avenues in the drinking water and wastewater field

This chapter focuses on career options available for drinking water and wastewater professionals. Without limiting the extent of scope, common job opportunities typically include as academic faculty, researcher, consultant, manager, engineer, designer, construction team member, system operator, laboratory staff, permitting authority staff, sales and marketing staff, manufacturing industry staff, and so on. Each job profile has specific educational requirements and involves peculiar duties or roles. Figure 3.1 shows potential job options in the water and wastewater field.

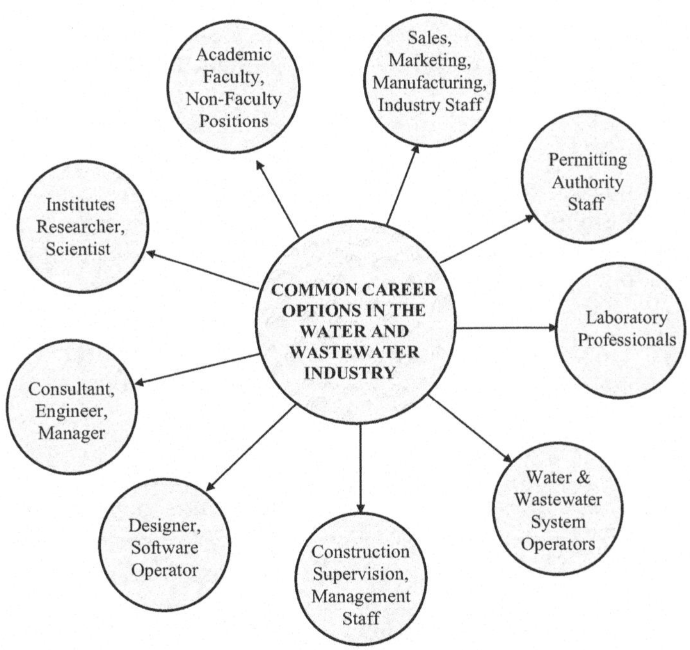

Figure 3.1 Career options for drinking water and wastewater professionals.

Company websites, recruiting sources, professional networking websites as well as several job boards in the drinking water and wastewater field can be referred to learn various jobs, titles and positions offered by organizations. Since job positions and titles are available on numerous sources worldwide, specific references are not listed for titles except for the university or college positions.

3.1 UNIVERSITY AND COLLEGE POSITIONS

Drinking water and wastewater professionals serving as faculty members are commonly appointed in *Environmental, Civil & Environmental* or *Chemical & Environmental* Departments in the universities or colleges. Master's degrees or Ph.D.s in Environment Engineering, Science or related areas are the common educational requirements for these positions. Various drinking water and wastewater related courses as well as other relevant environmental subjects are taught by these tutors. In addition, staff members may get involved in activities such as obtaining research grants, departmental funding and coordinating collaborative research work with other organizations or institutes. Faculties usually have graduate students or postdoc researchers to assist them in specialized research projects. Faculty staff present their research findings or studies in scientific journals, conferences or seminars. Faculty positions may be categorized as full-time, part-time, tenure track, non-tenure track, and so on. (Shamos, 2002; Duke University).

Depending upon the education and experience levels, professionals also get recruited for several non-faculty and administrative positions in the universities or colleges involving different roles. Although positions and titles can vary widely for states, regions and countries, the purpose of following list is to provide a general understanding of the various type of positions available in the academic area (Shamos, 2002; Duke University).

Examples of Common Academic Position Titles (Faculty and Non-Faculty)

- Adjunct Professor
- Assistant Consulting Professor
- Assistant Professor
- Assistant Research Professor
- Associate Adjunct Professor
- Associate Consulting Professor
- Associate Professor
- Associate Research Professor
- Consulting Professor
- Counselor
- Distinguished Professor
- Emeritus Assistant Professor
- Emeritus Professor
- Honorary Professor
- Instructor
- Lecturer
- Postdoctoral Associate
- Postdoctoral Fellow
- Postdoctoral Research Fellow
- Principal Lecturer
- Professor
- Reader
- Research Associate
- Research Associate Professor
- Research Professor
- Research Scholar
- Research Scientist
- Scholar in Residence
- Scientist
- Senior Fellow
- Senior Lecturer
- Senior Research Scholar
- Senior Research Scientist
- Visiting Professor

Some of the examples of administrative positions include Assistant Dean, Associate Dean, Dean, Program Director, Provost, Departmental Chair, Vice President, President, Senior Registrar and Registrar (Shamos, 2002).

3.2 RESEARCH INSTITUTE OR ORGANIZATION POSITIONS

Research organizations and institutes focused on drinking water or wastewater systems hire staff for research and development (R & D) activities. Such positions require specialized professionals, mostly Ph.D. or Master's degree candidates in the field. This is a great avenue by which researchers can get involved with innovation and development activities for newer technologies. Depending upon the job profile and responsibilities involved, the position titles can vary widely and here are just few examples:

- Research Engineer
- Research Scientist
- Research Manager
- Senior Research Manager
- Project Coordinator
- Director
- Executive Director
- President

3.3 CONSULTANTS AND ENGINEERING STAFF

This is more of an applied branch in which professionals are involved with live projects in companies for municipalities, boroughs, townships, cities, government organizations, private firms or developers involving drinking water or wastewater system infrastructures. Professionals working in these profiles often obtain certification or a license to practice in the area and are involved with planning, designing, modeling, cost estimation, permitting, bidding, construction and commissioning of full-scale projects (Lamont *et al.* 2007). Positions can vary from entry level engineer, consultant, surveyor, technician to the top executive positions such as company owner, president, vice-president, director, manager, and so on.

3.4 SOFTWARE OPERATORS, DESIGNERS AND DRAFTERS

Several types of computer-based software are used for the development of project designs, drawings, maps, system simulation and for modeling purposes (Benedetti *et al.* 2013; Gökkus, 1995; Li, 1992; Vanhooren *et al.* 2003). Companies frequently require professionals with expertise in operating such software (e.g., Computer Aided Design, CAD) for drinking water and wastewater projects. Such experts assist with the project task, such as, with converting survey data into drawings, developing hydraulic profiles, project layouts, equipment detailed drawings, preparing maps of wastewater conveyance or drinking water distribution network systems and various other applications. Unique courses and academic degrees are offered by universities and institutes focused on gaining expertise in these specialized software. Hence, designers, software system operators or programmers play a vital role in project teams.

3.5 CONSTRUCTION, SUPERVISION AND MANAGEMENT STAFF

Typically this type of profession involves field work and overseeing projects involved with the construction of utility pipelines, pump stations, treatment plants, pilot-scale studies and other components of drinking water or wastewater infrastructures. Construction and management experts need to handle several responsibilities such as managing project schedule, budgets, regulatory compliance, traffic management, mobilization and demobilization, inter-departmental coordination, permits, earthwork, equipment installation, and so on. (Commonwealth of Massachusetts, 2006; Lamont & Young, 2007; WSDOT, 2007,

2009). Professionals could also get involved with commissioning, start-up activities and initial operations of the systems. Some of the common examples of construction staff positions include the following:

- Construction Worker
- Construction Inspector
- Equipment Operator
- Construction Manager
- Project Supervisor
- Construction Technician
- Environmental Engineer
- Crew Leader

3.6 WATER OR WASTEWATER SYSTEM OPERATORS

System operators are heavily involved with the activities related to operations, maintenance and management of drinking water or wastewater systems, including collection, conveyance or distribution networks and treatment plants (Ohio EPA, 2013; US EPA, 2006). The operators are specialists who know the unit operations and treatment processes very well. These experts handle and manage water or wastewater equipment such as pumps, screens, grit systems, sedimentation, biological treatment units, chemicals handling, sludge management, storage tanks, booster systems, and so on. Following are some of the common job positions related to water or wastewater system operations:

- Treatment Plant Manager
- Assistant Manager
- Utility Supervisor
- Laboratory Supervisor
- Water or Wastewater Engineer
- Water or Wastewater Specialist
- Treatment Plant Superintendent
- Process Controls Specialist
- Plant Operator/Chief Operator
- Maintenance Crew
- Service Attendant
- Technician

3.7 LABORATORY PROFESSIONALS

These professionals perform laboratory duties such as sample collection, storage and analysis of drinking water or wastewater parameters. Staff members might also perform laboratory scale studies or bench-scale testing. Job profile can be a mixture of laboratory as well as field work. Staff are adept at handling laboratory specialized equipment such as a gas chromatograph (GC), mass spectrometer (MS), and so on. Water analysis typically includes parameters such as pH, suspended solids, biochemical oxygen demand (BOD), organics, volatile fatty acids, nitrogen, phosphorus, and so on. (Oregon DEQ, 2013; WS DOE, 2010). Some of the common job profiles in water or wastewater laboratories are listed below:

- Laboratory Director
- Laboratory Manager
- Program Supervisor

- Laboratory Analyst
- Quality Assurance Officer
- Lead Chemist
- Environmental Specialist
- Laboratory Technician
- Field Officer

3.8 PERMITTING AUTHORITY STAFF

The majority of the water or wastewater projects require approval from federal, state or local authorities for their execution. The permitting authority staff plays a key role in reviewing projects, understanding their implications on other utilities, the surrounding environment and community prior to approval. A project's compliance with regulatory programs makes it easier to ensure that the project will not have a significant environmental impact on the areas subject to regulation (Lamont & Young, 2007). Hence, the regulatory staff needs to be knowledgeable in the subject and well updated with current regulations and standards. Here are common examples of permitting authority staff members:

- Program Director
- Regional Coordinator
- Program Manager
- Administrator
- Environmental Specialist
- Environmental Scientist
- Environmental Engineer
- Water or Wastewater Engineer

3.9 MANUFACTURING STAFF

Several national and international companies are involved with the manufacturing of process equipment, laboratory materials, chemicals, pilot-scale plants, full-scale systems and other water and wastewater related components. Such companies hire environmental professionals to assist, oversee, test and manage the manufacturing process for drinking water or wastewater equipment.

3.10 UTILITY COMPANY STAFF

Private or government water utility companies are usually responsible for treatment and distribution of water within the city, township, borough or municipality. Such organizations need specialist staff to manage complex water systems and hire environmental professionals on a need basis. Sewer departments also hire staff to oversee the systems.

3.11 SALES AND MARKETING STAFF

Several manufacturing, consulting and design companies hire representatives to promote and sell their services and products. Such experts are well acquainted with the company's offerings and the process know-how they represent and help with business development and acquiring new projects. Common examples of sales professionals are as listed below:

- Inside Sales Representative
- Outside Sales Representative

- Regional Sales Engineer
- Technical Sales Engineer
- Water or Wastewater Sales Specialist
- Sales Manager
- Business Development Manager
- Marketing Manager
- Territory Development Manager
- Technical Specialist
- Area Manager
- Municipal Sales Manager
- Field Sales Representative
- Application Engineer

3.12 INDUSTRY STAFF

Chemical, mechanical, metallurgical, food and other manufacturing industries generate wastewater. Such industry discharges carry peculiar pollutants and need to be pre-treated prior to discharging into community sewers. Wastewater pretreatment processes are needed so that industrial wastewater pollutants do not cause interference with sewage treatment plant operations (OR DEQ). Also, drinking and process water demands are very high in processing industries. Such industries need to hire environmental, water or wastewater professionals to assist with managing water demands, wastewater pretreatment systems and ensure permit compliance.

3.13 JOB AFFILIATION ORGANIZATIONS

With various avenues available for getting involved in the water and wastewater related work, job openings can be available in a wide range of institutes and organizations. Here is a partial list of the types of organizations that could potentially hire professionals in the field.

- Government offices or organizations (Federal, State or Local)
- Private companies (consulting, engineering, design, utilities, developers, etc.)
- Manufacturers (equipment, lab materials, chemicals, etc.)
- Processing industries (chemicals, mechanical, metallurgical, etc.)
- Utility Companies or organizations (water and wastewater)
- Research institutes (Water and wastewater research)
- Universities and colleges (state, local, private, others)
- International organizations (United Nations, WHO, IWA, etc.)
- Non-profit organizations (water associations, federations, environmental groups, etc.)

Chapter 4

Professional development overview

Chapter 4 discusses professional development in the drinking water and wastewater industry. Sections included here provide guidance and key inputs for various stages of the profession. In general, the 'Student Phase' and 'Internship Phase' sections are important reads for students and newcomers, whereas the 'Entry-level Professional' section is useful for recent graduates with degrees, diplomas or certificates who have initiated the job application process or have recently entered the drinking water or wastewater profession. For professionals already in the business, 'Mid-level Professional' and 'Established Practitioner' are pertinent sections. Overall, all the chapters are relevant for every drinking water and wastewater professional at some point and one can use them for guiding newcomers looking for their direction and supervision as well. Since the professional journey can be significantly different for individuals and designations or titles can vary from company to company, general terminologies have been used for discussions. Figure 4.1 provides a brief overview of the professional development process.

4.1 STUDENT-PHASE

Although useful at every professional stage, the best use of this book to its fullest extent can be made by prospective students. Still in the molding stage of the career, the student phase is marked by enthusiasm, energy, inquisitiveness, curiosity, keen thirst for knowledge and big dreams. What really lacks is the experience and guidance in the right direction. Hence, educational institutions, parents and mentors play a vital role in shaping up the individuals.

To pursue a college or university education, students have various options available in the environmental and particularly drinking water or wastewater area. Many established institutions and universities worldwide offer diplomas, certificates, graduate-level and post-graduate level degree courses with a wide range of curriculum and subjects. Along with regular course work, higher studies also include seminars, projects, thesis or dissertation on specific topics (AACU, 2007; CRA-W, 2000). During this graduation phase, professors and teachers provide the necessary guidance to the students.

Thorough subject understanding and an extensive knowledge base gained during the educational period serve as the technical foundation for future employment. Additionally, students should make best use of available resources such as the internet, library books, social media, newspapers, television, association memberships, and so on, to keep updated with the recent trends and happenings in the drinking water and wastewater field.

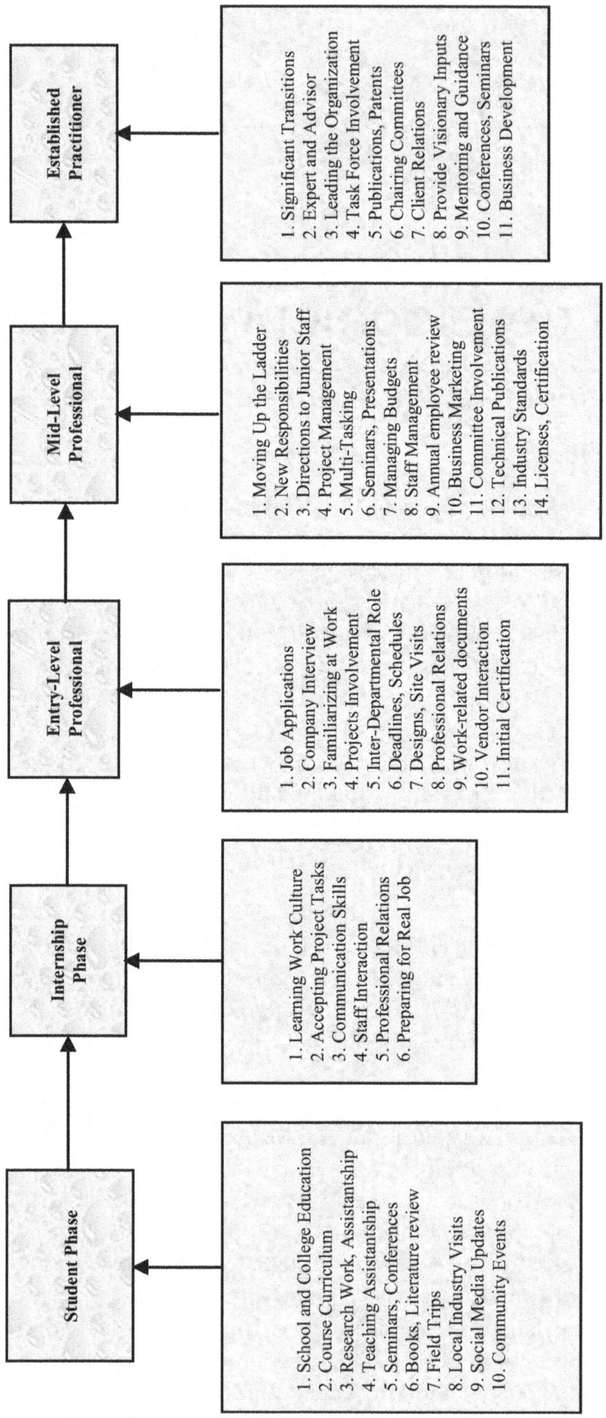

Figure 4.1 Overview of professional development process in water/wastewater industry.

Starting at school-level and continuing to successively higher education levels, students should prepare themselves for twenty-first-century challenges by gaining intellectual and practical skills, realizing personal and social responsibilities as well as understanding human culture and the surrounding physical and natural world (AACU, 2007).

4.1.1 Pursuing educational program

For students, it is important to understand the academic standards offered at a chosen school and the actual process required to obtain the degree (CRA-W, 2000). Before selecting a specific institution and educational program, students should thoroughly review factors such as degree of commitment involved, courses available in specific areas of interest, fee structures, institute location, faculty and staff reputation, credibility, campus jobs, feedback from past students, alumni, and so on, which are discussed in detail in the following paragraphs:

- **Commitment Involved** – Students should be completely aware of the commitment involved during educational studies prior to opting for a specific program. In general, diploma or certificate involves lesser coursework and are of shorter duration, whereas associate, bachelor's, master's or doctorate degrees involve progressively in-depth studies. Depending upon the institutions, educational programs may be available full-time, part-time, by correspondence or online (NSB, 2014; The Sloan Consortium, 2005). In addition to courses, higher studies also involve projects, seminars, dissertations or theses (AACU, 2007; CRA-W, 2000). With financial aspects involved, prospective students should understand educational obligations beforehand to avoid any time-schedule conflicts or reasons for leaving a program midway.
- **Preparing University and College Database** – As a starting point, students should start gathering information about potential environmental institutions such as establishment year, history, annual enrollment, acceptance rates, degree programs offered, achievements, specialized research areas, publications, institution's standing in the field, education fees, and so on. Developing an extensive database will help students to review and compare various educational programs available and identify a particular institution for pursuing studies in the environmental, and especially water or wastewater field.
- **Areas of Interest** – It is very important for aspirants to ensure that the courses offered at the potential institution match their areas of interest. This will have a vital impact on the quality of education gained, knowledge grasped and even sometimes grades obtained by the graduates. Hence, background work is necessary. After narrowing down educational programs, one should refer to websites, brochures and other available resources to review in detail the courses offered at the institute, areas of interests of lecturers and professors, ongoing departmental research work, funding received and projects completed by recent graduates. This homework would go a long way in successfully selecting the institution.
- **Credibility of Institution** – The institution at which students pursue their graduate studies is important, since the reputation of the institution can add to or subtract from the value of an individual's credentials upon completion (CRA-W, 2000). The background database developed should come in handy when analyzing the credibility and recognition of the institution by reviewing historical data, accreditations received, students' enrollment, awards, recognitions, college rankings, alumni networks, and so on. With this information, students will get a good understanding of the college they are planning to attend and what it has to offer.
- **Faculty and Staff Reputation** – College or university faculty members play a vital role in shaping up students and hence one should study the reputation and standing of lecturers, professors and

staff members at the institutions and within the environmental field. Often, college or university websites provide profiles of faculty members with details about their educational background, areas of work, published papers, conferences attended, awards received, and so on. This information is a good indicator of staff quality and their impacts on student education. Contacting faculty members working in the specific area of interest and requesting further program details can sometimes be quite effective with the institute selection process.

- **Location** – Colleges or universities of interest can be located in the same city, a different city, a different state, or even in another country. Depending upon the keenness for educational studies, available resources and ultimate professional goals, a decision can be made while selecting the location of educational program. Many students prefer to stay in the same or nearby cities with which they are well acquainted, or where most of their family and friends live. Some can opt for degrees in metros or other cities in the same or different states in order to experience a new atmosphere or study culture or to pursue an education at a higher ranking university. If the university or college education is highly regarded in the environmental or water and wastewater field, students can also plan to pursue education abroad.
- **Job Recruitment** – Various companies spend a great deal of money on recruitment. In many universities, recruiters may come to give informative sessions to select groups of students (AGU, 2001). Depending upon the quality of education, a college's ranking and faculty reputation amongst peers in the water and wastewater business, companies approach colleges to recruit fresh graduates. Institutions may occasionally publish statistical data on students recruited every year through campus job interview programs. Knowledge of such information can be a boost to pursue a specific program.
- **Past Graduates Feedback** – Any known past graduates of the institution can provide useful feedback about quality of education received, professors' attitude towards students, amenities offered, social gatherings held, personal experience with job search, reputation of college graduates amongst drinking water and wastewater industry members, and so on. A good feedback can be critical in a successful selection process.
- **Alumni Network** – A strong network of college alumni, their continued interest with the institution's developmental activities and personal involvement are signs of a good institute. Alumni typically have an emotional connection with their college, campus, programs, faculty and classmates. Good alumni are instrumental in developing a college's reputation within the industry and eventually with providing job opportunities for deserving graduates. Some university career centers develop a list of alumni willing to serve as contact points about various fields and students may access an institution's alumni database (AGU, 2001).
- **Financial Aid** – One key factor that students and their parents come across is the educational fee structure. With education fees and related costs getting higher every year, it has become an important consideration while selecting programs. Proper financial planning is needed that may include identifying various sources of funds as well as listing mandatory and discretionary expenses (NCREL, 1999). Some basic options for covering financial expenses include scholarships, loans, grants and work-study programs. The funding for these sources may come from universities, governments, businesses, private groups or companies (CRA-W, 2000; NEFE, 2011).
- **Student Visa** – Students pursuing studies abroad often need relevant visas (AGU, 2001). Hence, aspirants should get conversant with the formalities, documentations and requirements related to obtaining a student visa. Without a visa, other efforts towards obtaining admission or scholarship won't be fruitful. Such preparations should be made in parallel with pursuing the application and admission process.

- **School Applications Process** – The application process typically consists of requesting application materials, taking the required exams, asking for recommendation letters to be sent on a student's behalf, completing and submitting the application, sending letters of acceptance or rejection after receiving admission to the selected program (CRA-W, 2000).

4.1.2 Educational finances, fellowships and assistantships

- **Education Fees and Related Expenses** – These might include tuition fees, boarding costs, books, stationery, health insurance, personal expenditures, travel, and so on. Also, fees might be higher in metro cities compared to smaller towns or if education is pursued abroad. Cost economics should be carefully worked out before opting for the studies.
- **Educational Loans** – Upon selecting a specific educational institute and reviewing the fees structure, often acquiring a student education loan is a fruitful option. The loans can be potentially obtained from banks, government, private institutes, and so on, and there are many financial institutions supporting student education in different countries. Students should make sure to understand loan structures offered, associated interest rates, payment procedures, penalties, constraints, and so on, so that they are well aware of their financial status and can review future re-payment options (NEFE, 2011).
- **Fellowships and Assistantships** – Although obtaining a student loan is an option, pursuing financial aid might significantly reduce the monetary burdens on students and their parents. There are two basic types of graduate school aid available: fellowships and assistantships. Often, these scholarships require excellent grades or special talents and are mostly awarded on a competitive basis. Hence, students should strive to maintain higher grades and standing in the previous schooling years and show exceptional abilities (CRA-W, 2000; NEFE, 2011).
    - **Fellowship** – A fellowship is a form of financial aid that is similar to a scholarship. It is a grant of money for which no work is required, and can cover all or part of tuition, and it may include an additional stipend for living expenses. Fellowships are awarded based on merit and may be offered by universities, but most are through organizations outside the university such as private industry, foundations, and the government. Many corporations that sponsor fellowships may also provide paid summer internships. There is usually no requirement to work for the company after graduation; however, a successful internship may result in an offer of permanent employment (CRA-W, 2000).
    - **Assistantships** – There are basically two types of assistantships – Research Assistantship (RA) and Teaching Assistantship (TA). RA is a form of financial aid in which the student is required to work, often related to the student's studies or areas of interest. RAs pay the student to assist a professor in an experiment or a research project. Whereas, TAs pay students to assist in a professor's course or teach low-level undergraduate courses (CRA-W, 2000).
- **On Campus Jobs** – Even if a partial or no scholarship is obtained during the studies, often universities post temporary on-campus jobs for students that can partially reduce financial burden and educational loans. Such jobs can help to earn while learning and sustain self-education to a certain extent. Some examples of on-campus jobs may include cafeteria, library, gym, tutoring, and so on (NEFE, 2011).
- **Off Campus Jobs** – Similar to on-campus jobs, in certain situations, students can also pursue temporary off-campus jobs while studying. This can also assist with reducing the financial liabilities of education. Some examples of off-campus jobs may include retail, restaurant, baby-sitting, and so on (NEFE, 2011).

Considering the importance of scholarships and competition involved, it is recommended that students should prepare themselves well in advance and maintain good credentials at school level. The following suggestions can assist with improving the chances of scholarships.

- **Grades** – As is self-evident, students should strive to obtain and maintain good grades during their schooling and pre-university education to make sure that they stand-out distinctly amongst other candidates and look exceptional in terms of their credentials. Scholarships may be given to students for various reasons (NEFE, 2011).
- **School Projects, Seminars** – If students aspire to get involved with the drinking water and wastewater industry profession, they should try to pursue school projects and seminars in similar subjects and devote significant time to environmental topics to expand knowledge about this field. If there are any trade shows or conferences related to water or wastewater systems in the area, students are encouraged to attend them.
- **Social Awareness** – Several media outlets such as news channels, newspapers, internet websites, and local organizations on an occasional basis provide announcements, organize community events and share environmental updates. Students should follow such developments to keep themselves aware of any ongoing or upcoming drinking water and wastewater infrastructures projects within the community. Also, there are occasions when certain types of scholarships are announced by organizations for specific areas of study and advertised in media outlets. Student should keep eyes open for announcements of this type and news.
- **Library and Books** – School or community libraries are the best sources for books related to the drinking water and wastewater fields. It is quite important to get conversant with certain terminologies to improve understanding of topics and increase interest in this technical area.

With organized background work, students will be well-prepared and know the exact areas of study they want to pursue, understand the financial constraints, and pursue an appropriate university or college to achieve their educational goals.

4.1.3 Curriculum and subjects

After successful admission to a desired education program in an environmental and particularly drinking water or wastewater discipline, one should be meticulous in selecting subjects and coursework. Colleges or universities might have a fixed curriculum whereas some institutions may provide flexibility for selecting courses. From a professional's point of view, students should try to ensure that the selected university, college or institute offers basic and detailed courses covering a wide range of topics for water and wastewater systems. Below is a partial list of water and wastewater related topics that can be helpful for students with developing a good understanding of the field. It is noted that the courses offered by institutions worldwide vary significantly. Hence, these are general guidelines only and not intended to endorse any specific topics, subjects or coursework. Students are encouraged to work on details of their particular educational program with the college and faculty members. Also, no specific references have been listed here as the course subjects information is available widely on the websites of several universities, colleges and other educational sources internationally. Faculty members are the academic experts, typically involved with the development of specific coursework for water or wastewater subjects offered by respective institutions.

- Conventional and New Technologies
- Drinking Water Treatment
- Green Technologies
- Environmental Biology

- Environmental Biotechnology
- Environmental Chemistry
- Environmental Economics
- Environmental Geology
- Environmental Laws
- Engineering Design
- Engineering Mathematics
- Environmental Policies
- Fluid Dynamics
- Groundwater Systems
- Industrial Treatment Processes
- Instrumentation and Controls
- Limnology
- Plant Hydraulics
- Stoichiometry
- Sanitary Sewer Systems
- Sustainable Water Resources
- System Operations and Controls
- Utility Management
- Water and Wastewater Chemistry
- Water Distribution Systems
- Wastewater Treatment

Considering the scope of the drinking water and wastewater profession, course subjects similar to 'Drinking Water Treatment', 'Wastewater Treatment', 'Conventional and New Technologies', or 'Green Technologies' and related books would educate students about various unit operations and unit processes involved with the physical, chemical and biological treatment methods for producing high quality drinking water or generating treated wastewater effluent that can be safely discharged to the environment (Breach, 2012; Brepols, 2011; Libabher & Orozco-Jaramillo, 2012).

Studying 'Water and Wastewater Chemistry', 'Environmental Chemistry', similar subjects and associated literature would improve understanding of various chemicals and contaminants involved in the treatment processes, their reactions, kinetics, side-products, concentration and dosage requirements, mass balance, chemical properties and their applications such as corrosion, coagulation, sedimentation, precipitation, and so on (Hahn *et al.* 2004; Sonntag & Gunten, 2012; Tunay *et al.* 2010).

'Environmental Biotechnology', 'Environmental Biology', and similar subjects would focus on biological aspects of the systems. With several advances in the drinking water and wastewater systems and the inclination for 'greener technologies', focus has shifted towards natural, cost effective and energy efficient systems. These subjects would introduce students to concepts such as microbiology, bacteria types, growth rates, reaction rates, kinetics, chemical inhibitions, and so on (Ujang & Henze, 2003; van Haandel & van der Lubbe, 2012; Wuertz *et al.* 2008).

With issues related to underground contaminations, ground discharge of treated water, and so on, it is important to get hands-on knowledge of 'Groundwater Systems', 'Environmental Geology' and similar topics. Groundwater, aquifers, flow patterns, delineation of contamination plumes, treatment methods, locating underground utilities, storage tanks, and so on, are some of common topics encountered in the drinking water and wastewater field, and knowledge of these subjects would be an asset (Kazner *et al.* 2012; Milan *et al.* 2008).

Water lines and sewers are an integral part of community systems. Courses taken related to 'Sanitary Sewer Systems', 'Water Distribution Systems' and similar subjects would enlighten individuals regarding types of sewers, distribution networks, water pipes, sizes, capacities, pressure requirements, operations, maintenance, fire water demand, and so on (Alegre *et al.* 2006; Farley, 2003)

'Engineering Design', 'Fluid Dynamics', 'Plant Hydraulics', 'Engineering Mathematics' and similar subjects are very important from a professional point of view and for designing drinking water and wastewater systems, hydraulic structures, pump stations, process equipment, piping, and so on, as well as performing hydraulic calculations and generating profiles. These are some of the common tasks that water or wastewater professionals encounter as an engineer, designer or consultant and hence knowledge about these subjects would be critical (van Haandel & van der Lubbe, 2012).

All projects in the industry are governed by economics; hence the curriculum should include courses related to financial aspects of the industry. 'Environmental Economics' and similar subjects can help students understand project costs, interest rates, payback periods, life cycle cost analysis, financial planning, and so on (Merrett, 2007).

'Sustainable Water Resources', 'Environmental Policies' and similar topics for drinking water and wastewater systems in the curriculum will give students an upper hand in terms of up-to-date knowledge and ongoing trends in the industry before actually entering a job (Waite, 2010).

Courses related to 'Environmental Laws' and similar topics would be a useful addition to student's knowledge of local, state, federal and international environmental laws and regulations related to the drinking water and wastewater systems (Marchiso, 2000; Rouse, 2013).

4.1.4 Seminars and projects

Coursework completion is the basic requirement of a graduation program, however, seminars, projects and presentations provide additional dimensions to the studies. Collaborative projects provide experience in learning to work and solve problems in the company of others as well as sharpening an individual's understanding by listening to the insights of others, especially those with different backgrounds and experiences (AACU, 2007). An advanced degree gives students more flexibility in choosing projects and provides the expertise needed for increased levels of responsibility (CRA-W, 2000). The advantages, considerations and benefits of actively getting involved in these special tasks are summarized as follows:

- **Knowledge Application** – Seminars and projects are more applied by their nature as compared to classwork courses and may require extensive data collection, literature review, web search, field interviews, and so on. These tools if developed well can be beneficial to the real job scenario.
- **Teamwork** – Provides an opportunity to become a team player, coordinate with coworkers, share views and work duties, make adjustments and cooperate to achieve common project goals.
- **Decision Making** – Groups get actively engaged in the selection of an appropriate seminar or project topic, decide its scope of work, perform brainstorming to develop common interest topics, objectives, aims and approach, as well as review results and summarize conclusions. Such tasks inculcate a professional attitude towards the decision making process.
- **Computer Proficiency** – Technical presentations often involve preparation of slides, figures, graphs, charts, animation, and so on, and require knowledge of relevant computer programs. Hence, providing an excellent opportunity for students to develop computer skills and techniques for presenting their work (AGU, 2001).
- **Presentation Skills** – Exposure to presenting work in front of the class or group of students or faculty members can serve as a confidence building measure for professional practice. Time

management and maintenance of audience interest during presentations augment public speaking skills (AGU, 2001).

- **Defending Work** – A successful technical seminar or project not only involves good oratory skills but also practice at defending one's work and answering questions raised by the audience to their satisfaction. Hence, in addition to learning a particular subject in detail, students also learn how to defend their work.
- **Project Topics** – This is an excellent opportunity for students to explore and select drinking water or wastewater related current topics. A good literature survey can provide a basic understanding of present advances in the field. Topics such as green technologies, alternative energies, advances in treatment processes, innovative designs and similar other topics, if pursued, can be very helpful during job search and interviews as well.

4.1.5 Research and Teaching Assistantship

Classwork courses are the basic foundation and seminars or projects further add to communication and technical abilities, however Research Assistantships (RAs) or Teaching Assistantships (TAs) develop actual work related skills. Several important aspects related to RA and TA jobs during educational program are discussed below.

- **Working at College** – Both TAs and RAs are somewhat similar to actual professional work except that they involve lesser responsibilities and smaller tasks. Performing duties diligently, can help in the long run with professional development.
- **Teaching Assistantship** – TAs assist in a professor's course or teach low-level undergraduate courses. Assisting a professor may require grading problem sets or examinations, overseeing laboratory courses, teaching tutorial sessions related to the course, or office hours to explain problems relating to the course. TAs are usually awarded on a merit basis and many schools require students to TA at some point as training for academic positions (CRA-W, 2000). A student serving as a TA hence gains knowledge of what the teaching profession is all about, but in smaller doses.
- **Research Assistantship** – RAs assist a professor with experimental work or a research project. To be chosen as an RA is prestigious and offers several advantages. Students get directly associated with an ongoing research project, and may be able to formulate a thesis or dissertation topic as a result of the work. Students also may be able to conduct research for a thesis. Another advantage is that RAs may work with someone who may be well respected in the field. Published papers that result from the work would include the student's name associated with this respected individual. If these papers are presented at a conference, the RA may get the opportunity to present the work and make contact with others working in the field (CRA-W, 2000). With respect to water and wastewater fields, RA responsibilities may include equipment handling, experiment set-up, instrument operations, sampling and analysis, data interpretation, reports preparations, and so on. Many of the projects are live and time bound, so the students get experience similar to that of working in a research-based firm but again in smaller doses and with lesser responsibilities.
- **RA Job Opportunities** – Although an RA involves extensive research work in addition to the regular coursework, a research assistantship can lead to potential job opportunities. The research work involving collaboration and coordination with clients, can give students a platform for creating good relations with them and this can eventually lead to a job offer from the clients. Selection of the research topic involves key inputs from the guiding professor as well as student's interest in a particular area. Any innovative, ground-breaking research work and eventually good publications can provide name, fame and credibility to the student, faculty members and the institution.

4.1.6 Interaction with faculty members

Unless students have a very wise mentor, or a collection of advisors with experience in academia, industry, and other research environments, it is likely that they will be missing some important information about career options and opportunities in the field (AGU, 2001). Hence, mentoring and guidance at institutions are critical aspects for a student's development. As a learner, one should be receptive and good listener. There are several reasons for establishing good relations with professors who can assist individuals on a long-term basis:

- **Mentoring and Guidance** – A major figure in the life of the graduate student is the research advisor. The relationship that the student establishes with the advisor is vital to his or her successful completion of the graduate program (CRA-W, 2000). Also, professors and lecturers are the obvious mentors and guides for teaching subjects and theories, and resolving any doubts during the studies. Professors dedicated towards developing strong fundamental skills and technical knowledge in the students can have long lasting impacts on their careers.
- **Don't Be Shy** – With several students in the class, it is quite possible for some individuals to avoid asking questions or discussing doubts with professors in the class or after the class. To gain strong fundamental knowledge by the time graduation is completed, one should not be shy in asking queries and develop a habit of interacting with professors and having good technical conversations to resolve their problems. With appropriate and intelligent questions, an individual can win the professor's confidence and eventually develop a good long-term relationship.
- **References** – Contact references are typically included in the résumé for job applications with specifics including their name, job title, place of employment, relationship to you, full address, phone number, fax number, and e-mail (AGU, 2001). For pursuing jobs and higher studies, references are required to demonstrate one's technical abilities, subject knowledge and good personal character. Professors are the best referees for students and their good remarks can boost an individual's chances of securing admission for higher studies or a job position in a company. Also, a faculty's own industry contacts can potentially lead to job offers for the deserving graduating students.

4.1.7 Developing knowledge database

Overall, as the students keep busy working on courses, seminars, projects and possibly serving as RAs or TAs, they should continue developing their own 'knowledge database'. Gathering this information would be a part of personal development towards the drinking water or wastewater profession and may typically include class notes, study books, important articles, research papers, experts presentations attended, field notes, and so on. In this way, reference documents for the profession can be gathered in parallel while pursuing studies.

4.1.8 Social awareness of field

Study, database and personal notes should also be accompanied by improved social awareness of the drinking water and wastewater field. The awareness can be improved by following industry news, reading magazines, researching publications, joining professional organizations or subscribing to newsletters. Often, these publications are already subscribed to by the universities or institutions and students can refer to these resources from the institute's library. Keeping updated with recent happenings in the field and community can become an effective tool for job search.

4.1.9 'Special Remarks' Droplets

Well prepared graduates can get a decent start to the professional journey. With a good awareness about the profession, there will be comparatively fewer unknowns for the beginners. Also, education will be accompanied by confidence and a glimpse of practical knowledge to embark on the journey. Here are some 'Special Remarks' Droplets, which can provide useful tips to the graduating students.

- **Tip 1 – 'Out of the Box' Thinking Process:** it takes significant effort and devotion to develop an 'Out of the Box' thinking process. There are several established examples of researchers, industrialists and scientists who have gained name and fame for their unique approach, which is different from others in their field. Nurturing this habit can be a starting point for developing into an outstanding intellectual in the drinking water and wastewater field.
- **Tip 2 – Welcoming Feedbacks:** constructive tips and feedback from senior students, professors and other faculty or staff members should always be welcomed with an open mind. This receptive approach will help inculcate more ideas and develop a creative thinking process on a broader perspective.
- **Tip 3 – Starting from Home and Dorm:** all set for the drinking water and wastewater profession, one should start making efforts on a personal level in the home or dormitory with water conservation, proper waste management, understanding surrounding water and sewer systems and maintaining good sanitary conditions. These habits if advanced well can be a boost to one's thinking process and develop respect towards the profession.
- **Tip 4 – Inspirational Quotes:** collecting and following inspirational phrases and quotes by peers in the field can be quite motivating and eventually have a long lasting impact on the professional. Hence, one should keep gathering such phrases or idioms for self-reference.
- **Tip 5 – Time Management:** completing studies, working on seminars or projects and serving as an RA or TA can sometimes be demanding and difficult to manage. Proper time allocation and management can play a key role in minimizing work-pressure and at the same time helping on to enjoy the learning process.
- **Tip 6 – Being Focused:** student life is a developing stage and there can be many occasions when students can get distracted from their professional goals. It is important to remain focused, select friends and the study environment wisely and ensure that a high quality education is gained as a responsible graduate from the institute.
- **Tip 7 – Local Industries:** there is always a degree of contact between industry members and reputed institutes and universities in the nearby area and hence students can get good cooperation and encouragement for industrial site visits, or even for job opportunities. Hence, knowledge of the whereabouts of local industries, governments and municipalities that are involved in the drinking water and wastewater business can be beneficial.
- **Tip 8 – Personal Library:** if money allows, it is always a good idea to build a small personal library of technical books or publications on varied subjects within the field. The topics can cover a wide spectrum such as designing, fundamentals, chemistry, biology, tretment, operations, management, law and regulations, case-studies, and so on.

4.2 INTERNSHIP-PHASE

An internship is a unique form of experiential learning in which students get direct experience in a work setting, normally related to their career interests, and get the benefit of supervision and coaching from professionals in the field (AACU, 2007). An internship is a carefully monitored work or service experience in which a student has an intentional learning goal and reflects actively on what he or she is learning throughout the experience (NEDED).

Many companies promote internship programs. Businesses offering internship programs gain several advantages, and most importantly an opportunity to select and develop a company's future talent. Internships are also helpful for evaluating and screening students prior to making a full-time position offer, which can save the business money (NEDED).

A well-organized internship program involves plenty of training and supervision (NEDED). With internship programs, students benefit from on-job training, exposure to the work environment and practical knowledge gain, and develop confidence towards the profession. This is an excellent opportunity for students to be a part of the company while still pursuing a college degree program.

Internships can be typically of the following different types: project-related internship (assisting with specific short-term projects), summer internship (participating during flexible summer schedules), multi-year or semester internship (long-term programs working around student's class schedules and activities) and telecommuting internship (NEDED).

It should be realized that an internship is not always available to everyone and the individual getting this opportunity should take it seriously to make most out of it. Students pursuing internships learn the basics of the business, understand how to communicate with professionals and peers, are introduced to work culture and get prepared for actual jobs. Several aspects that should be considered during internships are reviewed in this section.

4.2.1 First step towards profession

An internship is the first step towards the profession and is a well suited platform for soon-to-be-graduates. The interns are provided with simpler tasks to perform, limited responsibilities and under the supervision of seniors. If approached from the right perspective, the internship can yield positive results. Also, it is an ideal way to begin the professional journey. Typically, internships are advertised by companies in the colleges or universities or are gained through contacts with professionals already working in the field.

These positions may be paid or unpaid. For businesses, there are several advantages for paying interns. Paying wages create an ownership in the program by both the business and intern. Unpaid interns may consider the position as a volunteer opportunity, hence potentially reducing their commitment or motivation. Also, advertising paid internships will likely increase student's interest in the position and boost potential candidate pool and improve chances of finding ideal candidate for the company (NEDED).

4.2.2 Basics of the business

To make an impact, one needs to learn the basics of the business and there can be no better time than learning this art during internship. Water and wastewater businesses typically run on financial, technical and managerial terms. Hence, as an intern, one needs to start getting acquainted with concepts such as budget, project scheduling, cost estimation, deadlines, meetings, designing, surveying, commissioning, and so on. These terminologies will become handy when a real job begins after graduation. Appendices A, B and C provide glossaries of basic terminologies used for water and wastewater systems, construction management projects and laboratory operations respectively.

4.2.3 Glimpse of professional experience

Internship provides an overview of an organization's functioning and operations. It provides a glimpse of experience that goes beyond just completing course work during graduation. As students return to the college from an internship to complete the course work, they start looking at class studies, seminars and projects from a completely different perspective and with a more practical approach. Broadening the horizons from short-term work experience only aids the development of high-quality graduate students.

4.2.4 Communication is the key

Company staff are the best resource for quickly learning the work environment. This can include administrative, technical, marketing, design or field staff. Each person and discipline has their own way of looking at work and projects. Hence, communication with the staff is the key for gathering the much desired knowledge. There is education and then there is practical experience. Internship is thus the best opportunity to learn from another's experience. One should try to get involved with company social events as it provides a good platform to interact with staff from various departments; during routine work days, one typically only interacts with selected members from project teams. Also, developing good rapport with the staff can eventually help with acquiring full-time employment within the company. Already established relations and known work culture can help individuals advance quickly in the firm.

During an internship, a mentor can prove to be an invaluable resource. A mentor will most surely enhance a student's experience and help them get the most out of their internship. A mentor can also serve as a reference when deciding whether or not the intern would be a good candidate to hire as an employee upon graduation (NEDED).

4.2.5 Preparations for a real job

With the final semester of graduation in progress, the next step is preparation for a real job. At this stage, one should start reviewing all the knowledge gained from studies, projects, seminars, presentations, conferences, RA or TA work and finally the internships. Realizing that job applications require résumé preparation, a cover letter, web search and networking, graduating students should start developing the necessary background work. Considering final exams and project workloads, a special time should be allotted to these activities in the last semesters; in this way, one is not bogged down by multi-tasking requirements. This preparation will ease the job search process and there could potentially be less downtime between graduation and the start date of the drinking water or wastewater professional job.

4.2.6 'Special Remarks' Droplets

Internship, the launch pad of a professional career, comes with certain responsibilities and benefits. Here are some 'Special Remarks' Droplets that can provide useful tips to the interns.

- **Tip 1 – Settle Down First:** Although a short time span, the interns should first settle themselves down in the working environment, as it is quite different from the college ambience. Don't be in a hurry to get results and prove yourself at work with the responsibilities allotted. Make sure to take time to understand the tasks, procedures involved and expected goals, and be receptive, calm and open to suggestions. This will minimize any mistakes in the first professional venture.
- **Tip 2 – Be a Team Player:** Remember that an internship is the starting phase and the job requires supervision by seniors. Be a team player, provide inputs, suggestions, results on given tasks and let seniors or managers take appropriate decisions. This way, staff members will develop confidence about the intern.
- **Tip 3 – Be Likable:** Make an effort to be likable to the staff members, be approachable and flexible with the work approach. Good friends and mentors can help interns to move in the right direction and get job opportunities as well.

- **Tip 4 – Respect Seniors:** For any conflicting opinions it is important for an intern to convey messages to the superiors in a respectful manner. This will avoid any possible friction and ensure that educational knowledge gained during college studies is well acknowledged by experienced personnel.

4.3 ENTRY-LEVEL PROFESSIONAL

After completion of course work, seminars, projects, theses, teaching or research assistantships and possibly internships in the company, the individual is all set to begin his or her professional career. The steps commonly involved are job search, interviews, appointment, training period followed by a drinking water or wastewater occupation. This stage is quite different from the internship as it is an actual transition from a well-nurtured college and student-life to a fully independent responsible professional environment. As the saying goes '*Well Begun is Half Done*', the individual should take the right steps to ensure a flying start to the career.

4.3.1 Job search options

The job search can sometimes be a tedious process and needs to be done smartly. There are several options for searching and pursuing career openings in the drinking water or wastewater industry and they are briefly discussed below:

- RA Collaboration Workplace
- Internship Company
- Faculty recommendations
- College or University Alumni
- Company Websites
- Professional Recruiters
- Online Networking Platforms
- Conferences and Seminars
- Personal Contacts

Often, research assistantship projects at college or university are collaborated with other institutions or organizations. During the assistantship, students get an opportunity to interact, coordinate and work closely with the staff from the collaborating organization. With extensive work, the individual can develop good rapport with the researchers, engineers or managers from the company. Upon completion of studies and depending upon the student's performance, the company can potentially provide a job offer. Hence, students should pursue this option if it fits the interest and goals of their selected career.

Internship companies are another obvious choice for pursuing a career. With the advantage of knowing the work culture and people in the company, the graduate might find job opportunities easier in this firm. Also, upon reviewing an individual's performance during internship, the company might feel comfortable offering a job and hiring the graduate.

College or university professors typically have good contacts in the drinking water and wastewater industry and often are approached by industry personnel asking for any suitable candidates for their firm. Being on good terms with professors, one can possibly get the chance to be contacted by the companies.

Alumni from the institutions are a good resource when pursuing professional positions. With several years of work experience and being an established professional, college alumni often look towards the graduating students from their college for potential job openings. With an emotional and personal connection with the institution, the alumni can have a significant interest in well qualified and high credential candidates, if available.

Now-a-days, in this 'internet savvy' world, several companies and organizations post job related advertisements and announcements on their webisites. These advertisements typically include information such as position title, job descriptions, professional hiring level, experience requirements, salary information, application deadlines, and so on. Upon careful review, the graduate can pick and choose ads, review company information such as services offered, projects completed, clients served, and so on, and select appropriate company openings for application.

Recruiters are the specialist firms that have a wide network of professional contacts with research, consulting, designing, construction, and other types of companies. Such companies tend to search for and hire candidates through these recruiters. Hence, contacting such professional recruiting firms, discussing available opportunities and selecting appropriate jobs for application can be an added option for graduates.

Online professional networking websites and job boards have become the prominent tools for connections and searching for opportunities. These networking sites are excellent tools for connecting with the peers and professionals in the drinking water and wastewater fields. Uploading one's profile gives the chance for others to review an individual's education, knowledge and professional achievements.

Attending conferences and seminars can provide good leads for the new graduates. One notable advantage of these events is the face-to-face meetings and personal communications. A direct dialogue is far more effective than one via telephone, email or other electronic media. If projected well, the students can impress company executive personnel and bag an opportunity to get interviewed by the firm.

Finally, an individual might have personal contacts or know someone in the field. These individuals can be in the designing, administrative, operations or management field. Reference or recommendation from these contacts can generate an interviewing opportunity for the graduate.

4.3.2 Job applications and interview

Once the job opportunities are narrowed down and specific company openings are identified in the drinking water or wastewater field, the job application process should be initiated. A cover letter and résumé or curriculum vitae are the basic documents used when applying for jobs. Employers review applications received from various candidates and short list potential candidates for the interview. The résumé and cover letter play a key role in successfully getting the interview call and hence should be taken seriously (AGU, 2001).

Résumé Preparations

Quite often, applicants do not have major experience with résumé or cover letter preparation. Several guidance books, publications and internet resources are available for preparing these important documents. Some of the basic components for résumé preparation are listed below (AGU, 2001):

- Education
- Projects, Dissertation or Thesis
- Professional Experience
- Academic or Professional Services
- Memberships or Professional Affiliations
- Grants and Funding
- Courses Taught, Students Advised
- Seminars and Conferences Attended
- Publications

- Awards, Fellowships, Honors, Scholarships
- References

Résumés should be prepared specifically for each job position applied. All personal and technical skills directly related to the opening should be highlighted. Résumés should avoid lengthy paragraphs or descriptions and possibly summarize them in terms of bullet points and short sentences.

Cover Letter Preparation

Prior to preparing the cover letter, applicants should carefully review the job description and compare with their career goals and objectives. They should ensure that the job profile matches with their expectations and interests. Many publications provide guidelines for preparing a good cover letter (AGU, 2001). Although cover letter submission might not be mandatory for all job applications, providing one shows candidate's seriousness towards the position as well as gives an opportunity to highlight applicant's credentials and experience that are directly relevant for the job position. Here are some useful thoughts about preparing the cover letter for water or wastewater jobs.

- Cover letter should be prepared specific to the job applied
- Letter should be concise, to the point and focused on job specific needs
- Candidate needs to identify educational qualifications that match the job position
- Skills and work experiences that are most relevant for the job should be highlighted
- Reason for applying to the position should be indicated
- Indications of how one can make positive contributions in the job position, if selected, should be provided
- Attention should be given to the flow of text and language used should be impactful

Job Interview

Following an intensive job search and submission of applications for relevant positions, successful candidates will be called for the job interview. This call is a good indication of company's interest in the candidate. Hence, adequate preparations should be made prior to the interview. Individuals should thoroughly review the company's background, history, profile, goals, leadership, key personnel, projects, service areas, and so on, before the interview. In addition, the candidate should read the job position profile meticulously and understand associated responsibilities and duties. This homework will make the candidate comfortable and confident towards facing the interview panel.

As an aspirant in drinking water or wastewater field, candidates should develop strong technical skills based on knowledge gained from educational coursework, projects, theses, research or teaching assistantships, industrial site visits, conferences, seminars, and so on. Understanding of water and wastewater systems, recent technological advancements, important equations and formulae, unit processes and operations, environmental laws and regulations, business structures, client types will prove handy for the interviews. Skill requirements may vary significantly for different jobs, profiles and companies.

Interviews help companies to evaluate if a candidate is a good fit for the business (NEDED). One should be relaxed, confident and fully prepared for this professional meeting. In addition to the strong technical qualifications and background, the individual should impress the interviewing panel with communication and organization skills as well as their logical problem solving approach (AGU, 2001). It is important for the candidate to convince how he or she is suitable for the position as well as how individual's knowledge and experience can be useful in successfully performing the duties.

4.3.3 Nervous mind at work

The job comes with a lot of responsibility, especially for the fresh graduate student. The environment they are used to during school and college days is quite different than that at the job. Hence, with the new working atmosphere, it is very common for the entry-level appointed staff to be excited as well as nervous at the same time. There are several ways to ease out this tension, get relaxed and be focused on the work. Some common tips that can help acclimatizing with the work environment are as discussed below:

- **Company Information** – Referring to the company website and related resources, one can develop a thorough understanding of the company goals, values, present status, ongoing projects, key personnel, staff members, and so on.
- **Office Area Familiarity** – Comfort in small things matters – so start getting used to the work area, desk, cabin, neighboring staff, dining area, parking, library, conference room, receptionist location, coffee area, lunch room, building layout, and so on.
- **Staff and Departments** – Since the majority of the tasks and projects in the drinking water or wastewater industry involve multiple departments, such as civil, environmental, mechanical, plumbing, electrical, and so on, one should start identifying staff with their divisions, their sitting locations, project involvement, and so on. This information will be helpful in approaching the right people for specific information.
- **Socializing** – Getting used to the office area and, upon introduction to the staff members from various departments, the next step is to start socializing with the professionals. This can be done as part of the project communication, meetings, seminars, during lunch time or at company social gatherings and events.
- **Rapport Building** – Focusing on work and keeping up with the deadlines, this is the time to build confidence amongst seniors and managers about one's professional and technical abilities. This would help one to settle down in the position and become an integral member of the company workforce.

4.3.4 Sticking to the basics

It is important to stick to the basics and build a professional career on the fundamentals learnt during the educational studies. As the individual starts getting involved with projects and becomes busier with time, one should keep the college and university books and notes handy for technical support and knowledge. Correct answers, right results and credible solutions can help win supervisor's confidence.

4.3.5 Profession based work documents

As responsibilities and involvements in the projects grow, the entry level professional should get accustomed to the basic work-related documents that need to be developed, reviewed or managed. These documents can vary with the type of job profile and profession selected in the drinking water and wastewater field. The technical documents may include research reports, manuscripts, policy documents, project documents (planning, designing, bidding, construction, commissioning and operations), laboratory datasheets, standards, regulatory permit documents, environmental laws, administrative papers, and so on.

- **Books** – Books are an integral part of the knowledge base in the drinking water and wastewater industry. They are creditable, in-depth and written by peers in the industry. Books are widely useful for many aspects of the job profile such as research, design, planning, and so on, as information related to the fundamentals, basic equations, chemistry, design basis, microbiology, technologies, and other related subjects is easily available via these references (Libabher & Orozco-Jaramillo, 2012; van Haandel & van der Lubbe, 2012).

- **Manuals and Standards** – These publications are different from books, have more applied contents and serve as a ready reference for practitioners. Since, publications of this type are typically not handled by individuals during their college period, professionals should spend some time to get used to them. Manuals and standards are usually published by local, state, national or international organizations and developed by experts, group of experts or task forces and are intended for a wider professional audience (Oregon DEQ, 2013; WS DOE, 2010).
- **Research Papers** – These publications are an integral part of the academic or research professions. Manuscripts published in journals are typically peer-reviewed and discuss original research findings or case studies by the professionals. With respect to the drinking water and wastewater field, specific topics can vary widely and may include new technologies, processes, water chemistry, biological systems, sludge management, hydraulics, simulation modeling, and so on. Papers usually follow a typical format involving background information, literature review, aim of study, materials and methods, research findings, tables & graphs, results, analysis, conclusions and suggestions followed by references. However, it is noted that the manuscript formats can vary from publisher to publisher (Ambulkar *et al.* 2011; Brandt *et al.* 2011; Olsson, 2007).
- **Product Information** – Product catalogues are commonly provided by manufacturers, sales representatives and vendors for specific equipment, technologies, instruments or particular products represented by the company or industry. These catalogues come in quite handy while comparing and selecting technologies, processes or equipment for the projects.

Without limiting the extent, the following is a brief description of common profession based documents:

- **Research-Based Profession Documents** – Researchers are mostly involved with the scientific research and development activities in the drinking water or wastewater areas. The projects may involve documents such as research proposals, technical reports, experimental data, laboratory-scale or pilot-scale study findings, scientific papers, conference presentations, laboratory manuals, and other relevant documents (Ambulkar *et al.* 2011; Brandt *et al.* 2011; Olsson, 2007; Oregon DEQ, 2013; Park *et al.* 2012; Tollan, 2002; WS DOE, 2010; Yan *et al.* 2010).
- **Academic Profession Documents** – Academicians, faculty members and other educational staff are typically involved with teaching courses in the environmental and especially water or wastewater areas. Education related documents may widely include course curriculum, quizzes, exams, grade sheets, attendance log-sheets, teaching notes, class presentations, technical posters, research papers, laboratory manuals, books, guidance documents, glossaries, and so on (Alaska DEC, 2013; CSU OWP; CRA-W, 2000; Libabher *et al.* 2012; NSB, 2014; The Sloan Consortium, 2005; van Haandel & van der Lubbe, 2012).
- **Planning-Based Reports and Documents** – Planning is the fundamental step towards initiation and implementation of projects. Types of planning documents can vary for different occupations, depend upon the organizational needs as well as project goals pursued. Development of planning documents may typically involve review of existing data (such as past reports, drawings, archives, maps, photographs, survey data, sketches, etc.), planning objectives, goals, analysis of various technical and non-technical aspects, financial planning, identification of immediate and future needs, and so on, and ultimately development of a visionary planning publication that becomes the basis for future projects and implementation strategies (Lombardo, 2004; Sousa *et al.* 2002; WHO & IRC, 2003).
- **Consulting Design Documents** – Professionals working in the design or engineering field are required to develop documents that serve as a solid foundation for the construction and installation of specific equipment, distribution pipe networks, treatment systems, tanks, reservoirs, technologies,

and so on. Depending upon the design stages (such as preliminary, primary, detailed or final), the extent and depth of technical involvements is decided. These may include spreadsheets, design calculations, system hydraulics, layouts, project drawings, specifications, permits, maps, bidding documents, technical reports, factsheets, formulae, handbooks, and so on (Lamont & Young, 2007; Ohio EPA, 2013; TN DEC, 2011; US EPA 1999a–c, 2000a–e, 2002a–c, 2003a,b, 2005a,b, 2013; USFA, 2008; van Haandel & van der Lubbe, 2012).

- **Construction Management Documents** – Construction activities involve implementation and execution of approved designs for drinking water and wastewater systems. These tasks may include upgrades to existing infrastructures or installation of new systems such as treatment plants, collection and distribution systems, new laboratories, and so on. The construction workers and managers need to maintain daily logs of such activities and develop documents like time schedules, milestones, workers and supervisor's logs, operations hours, equipment and inventories, permits, traffic control plans, budget, activity reports, progress reports, and so on, to keep track of the construction work (Lamont & Young, 2007; PPI Handbook, US EPA, 2000; WS DOT, 2007).
- **System Operation and Maintenance Documents** – Most of the routine activities for drinking water or wastewater infrastructures involve the operation and maintenance of systems. To perform and report the activities, documents such as daily log-sheets, reports, sampling, analysis, permit documents, O&M manuals, regulatory compliance records, violations, purchase orders for new equipment, and so on, are prepared and maintained (Gangl *et al.* 2007; US EPA, 2006; Wallace Scott, 2006).
- **Laboratory Operations Documents** – Laboratory professionals can be hired in research divisions of institutes, analytical laboratories, treatment plants and other relevant locations. Laboratory operators, technicians or other staff are required to manage laboratory activities as per the standards and regulations. Typical laboratory documents may include standard operating procedures, sampling and analysis data, instrument calibration procedures, equipment manuals, chemicals and equipment inventory list, safety regulations, purchase orders, permits, chain of custody documents, and so on (Oregon DEQ, 2013; Tollan, 2002; US EPA, 2005a,b; WS DOE, 2010).
- **Environmental Laws and Regulations Documents** – The majority of research, designing and implementation activities are governed and regulated by environment related laws and regulations. Hence, it is important to be aware of local, state and federal regulatory organizations and their published guidelines for water and wastewater systems. Many publications such as articles, papers, reports, regulations, and so on, are published worldwide discussing environmental laws, regulations and related aspects (Barry & LeBlanc, 2007; Marchiso, 2000; Rieu-Clarke, 2005; Schreiner & van Koppen, 2003).
- **Environmental Policy Documents** – Local, state or national governments as well as international organizations or experts develop environmental policies that can have influence on drinking water and wastewater businesses. These policies may discuss priority issues, initiatives, goals, incentives, funding, and so on. Professionals should keep updated with these policy documents (Brown *et al.* 2002; Rosemarin *et al.* 2008; Schreiner & van Koppen, 2003; UNICEF and WHO, 2012; WHO, 2008).
- **Administrative Documents** – Irrespective of the job profile selected, professionals are required to maintain administrative documents as a part of the profession. These may include work time-sheets, administrative correspondence, appraisal forms, vacation or leave logs, and so on.

Getting conversant with documents of this type beforehand can be very comforting for the new professionals and contibute to efficient working. Keeping one's own small library on the desk or cabin for reading and understanding each type of document, till one gets used to them, can be quite beneficial.

4.3.6 Inter-departmental coordination

For an entry level professional, one of the key aspects to quickly catching up with the speed of others in the company is to get used to inter-departmental coordination. Often the projects are fast paced and with a short budget requiring significant coordination within the team. Although managers or senior professionals may handle the majority of project coordination, the entry-level professionals are provided with responsibilities to work with other disciplines on certain tasks. Hence, good communication and rapport with other departments are important considerations.

4.3.7 Vendor correspondence

Vendor correspondence is a new concept that the entry level professionals get introduced to at work. As potential suppliers of equipment or services, professionals interact with vendors and manufacturers. These communications are a fruitful and valuable learning experience for the individuals. To sell the best products and services in a competitive environment, vendors provide brochures, drawings, datasheets, specifications, cost estimates, proposals for their products or services and are also willing to answer any technical questions or queries that the professionals might have. These interactions introduce one to lot of field photographs, videos, samples, and so on, about products that provide a completely new dimension to the understanding of drinking water and wastewater topics.

This is a good phase to learn the drinking water or wastewater unit processes and operations in details. Water and sewer pipes, pump stations, booster stations, televising videos, manholes, screening methods, grit removal processes, process tanks, new treatment technologies, sludge digestion processes, chemical feed systems, and so on, are some of the examples. Audiovisual documents or photographs received from vendors are an asset for the entry level professional to take next step towards obtaining in-depth knowledge.

4.3.8 Managing deadlines

With multi-tasking and work pressures, junior level professionals sometimes tend to get bogged down by the workload and could potentially miss the budget or important deadlines. Hence, one needs to understand that the workplace being a different environment than college where students get plenty of time to complete their submissions, one has to keep up with the speed, dates and time in order to finish tasks. It is imperative for a successful professional to get used to the deadlines and budgets. There are certain approaches that can assist with this time management process:

- **Multi-Tasking** – Keep a track of all the different ongoing and upcoming water or wastewater projects and tasks that the professional has to perform to satisfy client requirements.
- **Talk Beforehand** – If based on the current workload, individuals feel that upcoming tasks and deadlines are unrealistic, they should inform the manager right at the initiation or planning stage of the work allotment. Inability for timely updates could potentially leave a bad impression amongst seniors and they could lose faith in one's ability to keep up with the time schedule.
- **Time Management Gadgets** – With advances in internet technologies and electronic gadgets, several time management calendars, activity books, reminders, and so on, available in the market or even within the company should be used. Smart use of electronic tools can relieve professionals from constantly remembering the deadlines, meetings or scheduled events.
- **Holidays and Vacation Schedules** – Time sensitive work should take into account any upcoming holidays or personal vacations. Also, one should be aware of vacations taken by other staff members within the project team, if possible, to manage workload. Neglecting or forgetting this aspect can have a negative impact on meeting deadlines.

4.3.9 Initial certifications

For continued growth in the industry, certain countries have a requirement that professionals pursue industry related certifications or licenses as a part of their work profession. Although academic degrees are respected well in the industry, additional certifications and designations improve credibility for drinking water and wastewater services offered by the individuals. With industries becoming global and collaborative projects being pursued by national and international firms, many professionals are obtaining additional designations to advance their position and standing within the drinking water and wastewater community. Obtaining additional certifications and designations will not only improve one's respect amongst peers, but also help one to advance to senior positions and gain salary rises.

4.3.10 'Special Remarks' Droplets

Entry-level jobs can be demanding as they involve multi-tasking, specific expectations by seniors, establishment in the field, development of new professional relations, and so on. Here are some 'Special Remarks' Droplets that can provide useful tips for entry level professionals.

- **Tip 1 – Avoid Quick Conclusions:** As the professional's work commences, excitement and enthusiasm are quickly replaced by workloads and responsibilities. One starts interacting with already seasoned and experienced professionals. These interactions are quite different than those at family, friend or college level. So, don't rush to quick conclusions about someone's attitude or behavior towards you or your work. Don't be judgmental about others and give time for more interactions before forming an impression about others. This will help you to select the right friends and colleagues within the company and can help you develop long-term relations.
- **Tip 2 – Professional Tools:** As a researcher, planner, designer, surveyor or operator, there are several different types of tools and equipment used at work. These may include laboratory instruments, process monitoring tools, measuring equipment, drafting tools, surveying equipment, and so on. For foolproof output, the theory and operations of these tools should be well understood.
- **Tip 3 – Seminars and Conferences:** Be open and attentive to any announcements about seminars, conferences or lectures related to water or wastewater systems. Attending such events can improve understanding of the subject, expose one to new technologies and even provide a platform for networking and interaction with professionals working in other organizations.
- **Tip 4 – Project Site Visits:** Till now, majority of the knowledge or perception received by entry level professionals about drinking water or wastewater related processes is via pictures, photographs, descriptions, books, and so on, during college and during internship jobs. There are many unknowns and assumptions at this point which might not provide the complete vision while planning or designing projects. It is important to make site visits and look at the actual systems in operation, to interact with operational staff and other field professionals. This is bound to add a completely new dimension to the thinking process about these systems and how the real drinking water or wastewater world works.
- **Tip 5 – Organizations Membership:** Already in the profession, individuals should try to expand their horizon of knowledge and networking. There are several local, national and international organizations in the environmental field that require volunteers and professionals for the advancement of drinking water or wastewater knowledge. Obtain membership in such organizations and try to get involved with young professionals or technical committees within the institutions. It will provide a platform to interact with peers from other firms that are working towards betterment of the profession by supporting and devoting their time for these activities.

- **Tip 6 – Balancing the Transition Phase:** The entry level phase is often marked by developments at a personal level as well, such as moving to different city or countries for jobs, away from parents and families, getting married, and so on. These changes are also accompanied by a new job environment and responsibilities with salary in hand. To minimize any mistakes at this juncture, be patient and don't get overwhelmed or overexcited by multiple transformations in your life.

4.4 MID-LEVEL PROFESSIONALS

Moving up the ladder from entry or junior level to mid-level positions is not an easy task and requires sustained efforts and significant drinking water or wastewater experience. As mentioned earlier, company hierarchy structures vary widely from region to region, depend upon the profession (such as consulting, designing, research, construction, operations, etc.) and even the organization type (i.e., government, private, international, non-profit, etc.). With this, the mid-level position titles can be significantly different worldwide. The 'mid-level professional' term is used in this book to generally represent seniors or mid-managerial professionals.

The mid-level phase is marked by greater responsibilities, more coordination efforts and higher sets of goals. Some of the salient characteristics associated with the a mid-level professional's role include project management, multi-tasking, inter-departmental coordination, client correspondence, hiring new staff, gaining professional certifications, staff evaluation and involvement with key organizations or committees.

Project managers provide a very active leadership and are responsible for all phases of the project, from project development through construction, start-up and operation (Lamont & Young, 2007). There are several books by well-known authors related to the management and business administration aspects with in-depth analysis and guidelines. Hence, discussions in this section are limited to the aspects relevant to the context of the book and as appropriate for describing the drinking water and wastewater profession.

4.4.1 Project management for infrastructure projects

Successful implementation of a project involves a multitude of issues besides those of applied science, including communications, management, and leadership skills (Lamont & Young, 2007). Completing specific tasks at entry-level or in a junior position is totally different from managing projects. A successful manager needs to develop thorough understanding of the project and company goals, and inculcate creative methods for team-building and project management. Following are some of the basic tasks for managing drinking water and wastewater infrastructure projects:

- Review proposals, budget, scope, schedules thoroughly prior to accepting or signing project contract
- Check available staff, and their qualifications prior to initiating specific work
- Complexity of projects and permit requirements should be reviewed beforehand
- Traffic, weather and other factors affecting a project should be analyzed in detail
- Ensure that the project complies with required standards, regulations and laws
- Engage in regular client meetings and discuss scope or budget changes on a timely basis
- Regularly present to and update team members with project progress and status
- Perform routine quality checks of team outputs to ensure integrity of work
- Keep top management updated with project status and compliance with company goals
- Provide guidance to junior level staff and ensure they clearly understand project goals
- Attend conferences, seminars and other professional events for new business opportunities

4.4.2 Effective coordination and communications

Effective coordination and communication plays a vital role with a project's success and professionals should coordinate necessary project correspondence between the parties and company departments involved (Lamont & Young, 2007).

- **Parties Involved** – For efficient communication it is important to identify parties involved with the projects. These may include clients, contractors, vendors, regulatory authorities, developers, and so on. To ensure smooth operations, all the parties should be kept in the loop and updated regularly during the project life span. A list of all project participants should be developed and their contact information be made available to the project team (Lamont & Young, 2007).
- **Inter-Departmental Coordination** – Depending upon project scope, many departments within the company such as civil, electrical, mechanical, piping, computer programmers, and so on, may get involved. In such cases, appropriate managers or staff members of respective departments should be contacted, scope should be discussed and work coordinated efficiently.
- **Meetings** – For effective work, departmental and inter-departmental meetings are necessary. These meetings help to keep the team updated with the progress, identification of upcoming issues, staff and budget situations, technical staff management and getting everyone in agreement with respect to the deadlines and milestones of the project.
- **Communication Methods** – Emails, calls, faxes, transmittals and so on are all different ways to communicate during a project. The project manager should ensure that there is effective communication and that project team members are copied with pertinent correspondence and included in project meetings, conference calls, email lists, and other methods of communication (Lamont & Young, 2007).
- **Project Timelines** – Projects are quite often time sensitive and any lack of coordination or miscommunications can result in unnecessary delay in the completion of work and submission to the client. Hence, once the schedule and deadlines are finalized, they should be conveyed promptly to the relevant professionals of other departments.
- **Budget** – Just like the project timeline, budget plays an important role with the successful completion of a project. Thus, in addition to the schedule, budget planning should be prepared, conveyed to the departments involved and reviewed on a regular basis (Lamont & Young, 2007). Drinking water and wastewater project sponsors can be government offices, industries, private developers, international organizations, and so on, and have different budget related needs and constraints.
- **Flow of Information** – The inter-departmental information flow is quite critical for meeting deadlines and budgets. Delays, mismanagement or lack of data transfer can result in significant loss of time and excessive manpower requirements to complete the work and can potentially overshoot the project budget. Also, timely flow of information develops faith and confidence amongst other staff members regarding the project and is a key aspect for the team-building process. Available company communication platforms and related software should be used efficiently.

4.4.3 Non-technical duties

Apart from project management tasks, mid-level professionals in the drinking water or wastewater business may also be expected to perform several non-technical duties and overview staff activities including the following:

- Interviewing and hiring new staff members
- Handling employee performance reviews

- Managing staff vacations and workforce availability
- Involvement in company budget planning
- Coordinating social gatherings with administrative staff
- Meetings and lunch/dinner with clients
- Marketing and presentations at conferences and seminars

With multiple activities on the desk to perform, mid-level professionals need to maintain good rapport with colleagues, junior staff and senior management. Also, one's behavior and principles should help develop confidence and leadership authority within the company.

4.4.4 Client correspondence

As a mid-level professional, most of the time, client correspondence and coordination becomes a major part of the profession. The drinking water or wastewater industries, like other businesses are driven by economics and major investors in the infrastructure projects mainly include governments, townships, boroughs, industries, developers, regulatory authorities, research organizations, and so on. Being a client contact, the professional should be well aware of client type, its requirements, mode of operations, way of communication, relevant contact personnel, their understanding of technical knowledge, and so on. This background will help professionals realize what, how and where to discuss the specifics to ensure project integrity, high quality of work, excellent professional services as well as minimizing mis-communications, avoiding conflicts of interests and developing good client relations.

4.4.5 Learning industry reference standards

Progressing higher in the profession, mid-level professionals are loaded with several project responsibilities and obligations. At this stage of a career, it is important that the individuals are well aware of important industry standards considering the liabilities and insurances involved with the projects. Industry reference standards are used as effective tools for specifying and selecting project equipment, laboratory instruments, process control systems, treatment technologies, structural, civil or mechanical materials, and so on, and are published by numerous organizations worldwide.

4.4.6 Professional designations

A variety of professionals and technicians (such as planners, regulators, engineers, designers, installers, operators, pumpers, and inspectors) get involved in different aspects of the water or wastewater projects. Training, along with certification or registration, provides system owners and users with competent service providers and 'raises the bar' in promoting professionalism among the industry (US EPA Manual, 2005a). Many companies give importance to educational degrees and work experience for moving up the professional ladder. Whereas, in certain instances additional certifications, licenses or designations are required in the drinking water and wastewater field to achieve higher positions and boost careers. Here are some examples:

- Professional Certifications
- Professional Licenses
- Operator Certifications
- Scientific Designations
- Environmental Registrations
- Management Degrees

Requirements for obtaining such designations, certifications or licenses can vary significantly for states, regions and countries throughout the world and type of job profile pursued. The purpose of introducing this concept is to identify additional advancement avenues for professional development in the field. Individuals should check local practices related to this profession. Many national and international organizations provide such recognitions and often involve certain educational requirements, work experience, test examinations and interviews for successfully obtaining them.

4.4.7 Key organizations involvement

Delivering best management services in the company is an important role, however, to move beyond the company circle and get acknowledged by the outside drinking water and wastewater business community, there are many avenues available for the professionals. Several national and international organizations provide a common platform for professional involvement and moving the industry forward. Joining committees, task forces and specialized events would not only provide an opportunity to develop newer professional contacts but also expand the knowledge base and get more industry updates with firsthand information from leaders in the field.

4.4.8 'Special Remarks' Droplets

One might wonder what makes some professionals 'standout' distinctly amongst other mid-level professionals and managers while performing the same roles with similar responsibilities. There are few simple considerations and techniques that can help establish individuals as superior amongst others in the industry.

- **Tip 1 – Respecting Coworkers:** Although this looks like a simple philosophical term, it does go a long way in establishing superiority within the profession. Respecting others will help earn respect from colleagues as well. Respect leads to recognition.
- **Tip 2 – Leading from the Front:** With their working qualities and ethics, professionals should set an example for other colleagues and junior staff to follow and be inspired by. This would build confidence about their team leadership.
- **Tip 3 – Supportive of Staff:** Be supportive of staff and try to resolve any conflicting issues to move the team forward in the correct direction towards project goals. Often, newcomers need help and guidance. Keeping faith and providing support can develop good team spirit.
- **Tip 4 – First Come First Served:** Professionals should be fair and neutral towards other team members. Unless there are any immediate deadlines or instant needs, attempts should be made to resolve issues on a first come first serve basis.
- **Tip 5 – Ready for Unexpected Situations:** There could be instances when things might not appear fair, logical or ethical. Also, egos, group-ism, and many such factors can surround the work culture. It is important that one should follow strong ethics and work within the principles that they have developed over years. Exceptional individuals get recognized by the way they approach and resolve situations.
- **Tip 6 – Guiding Staff Members:** Managers are always approached by juniors for work related, or, occasionally, personal suggestions or guidance. This is the best opportunity to build one's image amongst the staff member by providing logical, creative, ethical and useful guidance.
- **Tip 7 – Trying Times:** Although the aim may be noble, the approach pure and goals clear, there could be trying times when things don't work the way we want. One should be prepared to handle all the situations with a fighting spirit, keeping the interests of society in mind, and serving as a responsible professional of the drinking water and wastewater industry.

- **Tip 8 – Good Listener:** Good managers or professionals are good listeners. Everyone is unique and comes with a unique set of talents. Never take others' suggestions lightly. Be a good listener and respect suggestions by staff, even if they are by juniors. A successful manager acknowledges others' talents and ensures that constructive inputs are implemented.
- **Tip 9 – Understand Top Management:** Mid-level professionals are answerable to the top management and executives of companies and hence should develop a clear understanding of a company's vision and the results expected by the top authorities.

4.5 ESTABLISHED PRACTITIONERS

Practitioners go through major transitions in professional life by the time they establish themselves in the industry. Enriched with significant knowledge, expertise and being at the top of the field, established professionals enjoy special status in the drinking water and wastewater community. Practitioners are the leaders in their specialized areas and have the ability to navigate the future of this industry. Here are some of the salient features of being an established professional:

4.5.1 Significant transitions

Experience teaches a lot. There is quite a 'learning curve' starting from the student-phase all the way to the established practitioner's position. Education is slowly overshadowed by experience. Theoretical skills get replaced by real world practical expertise. There is a substantial knowledge base gained in terms of technical know-how, managerial skills, presentation skills, client relationship skills, and so on. Recognition in the field and respect by peers becomes the highlight for business practitioners. Quite a journey…!

4.5.2 Pillars: Where, when and how

Not only is it important to reach the top of the drinking water or wastewater profession, but it is also important where, when and how it is achieved. These three pillars shape up the practitioner's voyage and eventually surround him or her as an 'aura' in the form of respect, recognition, name and fame within the professional society. Here are a few thoughts regarding the three pillars:

(1) **Pillar 1 – Where:** The first pillar realizes where and under what situations the professional has achieved accomplishments. Factors may include an individual's background, financial conditions, city, country, school, college, university, work places, job profiles, companies, and so on. A practitioner's '*direct influence on the industry*' is a result of individual contributions, team involvements and interactions with teachers, professors, alumni, company colleagues, clients, staff members, juniors, and so on – people with whom the professional has worked closely in direct reference to the drinking water and wastewater field. Also, additional involvements in local, state, national or international committees, task forces, panels, and so on, or at other important platforms result in an '*extended influence on the industry*' with direct or indirect contributions.

(2) **Pillar 2 – When:** In a fast paced world, aspirants may desire to climb the professional ladder swiftly and attain an established practitioner's position at a relatively early stage of the career. We are aware of some well-known names in several professional fields (not limited to drinking water or wastewater), who are established experts at an impressively young age. Since individuals have their own career paths, struggles, and ups and downs in the profession, the age at which one starts developing this vision is really important if one is to take it all the way to becoming an expert. Timing and duration play a vital role in the establishment process and eventually impact an individual's '*professional aura*'.

(3) **Pillar 3 – How:** The third pillar overviews how this journey was carried out. Following strong ethics, developing sound principals and ensuring respectful behavior, individuals gain prestige and reputation from industry members. For a journey voyaged with noble thoughts and approach, the '*professional aura*' only gets bigger and better with time.

4.5.3 Types of established professionals

The definition of an established practitioner can vary considerably between industry members. Company ownership, executive positions, leadership qualities, monetary assets, experience, fame, knowledge, patents, publications, and so on, are some of the parameters for measuring success and expertise in the drinking water and wastewater business. Although the journey starts and moves progressively from the student phase, it should be realized that not everyone eventually owns or leads an organization. However, individuals in various levels of hierarchy attain noteworthy expertise in specific areas. Hence, an attempt is made to cover '*professionals with extraordinary abilities*' from various strata of the industry. Some of the examples are as follows:

- Company owners, partners, entrepreneurs
- Organization executives and managers
- Academicians and scientists
- Well-established engineers, designers, researchers or operators

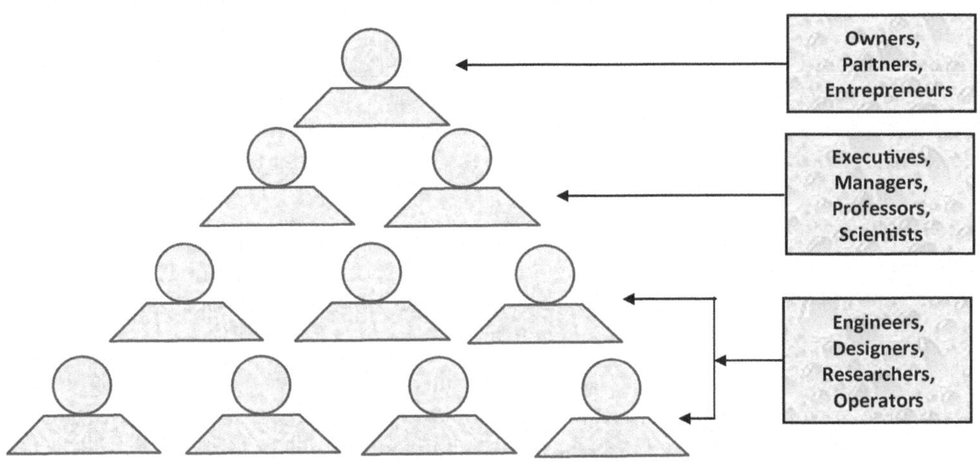

Figure 4.2 Established practitioners at different levels of workforce.

(A) Company Owners or Entrepreneurs

With their vision and experience, certain individuals start their own companies or businesses. These drinking water or wastewater firms may relate to different types of services such as manufacturing, sales, consulting, designing, research, laboratory, surveying, operations, construction, and so on. Owners are the 'visionary' members for their organizations. Also, there can be multiple owners or partners in the business. These individuals oversee the functioning of the firm from a professional, financial, managerial and legal point of view. Depending upon services and growth, companies hire managers, engineers, designers and researchers to execute projects. Owners typically involve themselves with company vision, client relations, business development, finances, future directions, and so on.

(B) Organization Executives and Managers

This is the second type of established practitioner's category and might include executives, managers, directors, presidents and similar positions. Such experts provide key inputs and directions towards company growth in terms of acquiring new projects, company vision, policy development, staff management, directing and guiding junior staff, so on. These practitioners may also get actively involved with national and international drinking water or wastewater organizations in key roles such as chairing committees, task forces, technical and advisory boards, and so on. Members also handle the responsibilities of inter-departmental coordination, presentations, client relations, and overall company management.

(C) Well-established Engineers, Designers, Researchers and Operators

Although serving at lower hierarchical positions such as engineers, surveyors, operators, designers, managers, and so on, these professionals make their mark with some extraordinary knowledge gain and contributions towards the industry. These achievements can be in terms of excellent performances, awards, patents, involvements with key professional organizations (committees, task forces, technical and editorial boards, etc.) and by authoring scientific or technical publications (like manuals, technical standards, books). With multiple engagements, these individuals get themselves recognized amongst the professional community on a wider scale and established as a key member of the industry.

(D) Academic Professors and Researchers

The next category of extraordinary professionals involves academic and research staff. Typically after completion of a master's degree or Ph.D. degree or post-doctoral research work, individuals get engaged in research-based or academic activities. With significant experience, publications, book authorship, serving on technical boards of scientific and research-based journals, groundbreaking research work, patents, and so on, individuals get acknowledged as established practitioners in the industry. Members play a key role with significant advancements in research, innovations and academics.

4.5.4 Mentoring and guidance

In any of the above mentioned practitioner categories, experts gain significant knowledge that can be used for mentoring and guiding others in specific areas. Here are some examples of individual profiles and knowledge base they can offer to others in the industry.

(A) Owners or entrepreneurs would typically have substantial knowledge about how companies are established and managed, hence they can provide key guidance about how to conceptualize, develop and advance a company, what are the technical, financial or legal matters involved, how to create new clients and expand business structures, and so on.

(B) Senior managerial and executive staff members will be able to provide key inputs about how to handle projects, develop client relationships, manage company staff, financial planning, policy decisions, inter-departmental coordination, company growth, committee involvements, and so on.

(C) Established engineers, operators, researchers and surveyors will be able to provide guidance for technical aspects related to drinking water or wastewater subjects including technological advances, the planning process, designing, system operations, analysis, case studies, manuals, standards, and so on.

(D) Researchers and academicians can share their expertise related to teaching, education, research and development activities. The high quality of scientific advances can result in ground-breaking work in the area and these established mentors can provide crucial guidance in the research profession.

Each type of established practitioner has his or her own peculiar strengths and can contribute or offer unique knowledgebase to the professional community.

4.5.5 'Special Remarks' Droplets

Established professionals carry the highest responsibilities towards the industry and there are certain aspects that should be considered at this juncture of the occupation:

- **Tip 1 – Social Contributions:** One of the most regarded and necessary contributions expected from established professionals is towards society. By this time, experts have moved well ahead in the 'Learning Curve' of the profession and it's time to give back to society. Contributions can be in the form of informative lectures, financial donations, advisory roles, volunteering activities, and so on.
- **Tip 2 – Technical Publications:** Professionals should make contributions by writing books, authoring policy documents, manuals, standards, factsheets and other important publications that can be referred to and used by the budding professional community. This is a good way to share experiences that the established practitioners have gained over their professional life span.
- **Tip 3 – Journals and Conference Reviews:** Authoring leading publications is one aspect whereas reviewing and commenting on other expert's work is another. There are many such platforms that provide professionals an opportunity to judge the work of others and make valuable contributions by providing constructive comments for improving quality and outputs. As a peer-reviewer, a professional can serve as a journal reviewer or conference technical referee, or can be an expert providing comments on international reports developed by global organizations related to drinking water and wastewater.
- **Tip 4 – Creating New Organizations or Societies:** Drinking water or wastewater related non-profit organizations provide a great platform for professionals to come together and share experiences from varied job profiles and institutes. Although many such organizations are already active in the field, if there are any new or innovative ideas for which no other organizations are operational or focused at present time in the region, this would be a great opportunity for established practitioners to start new ones.
- **Tip 5 – Professional Networking:** Several internet-based options are available for socializing and networking with peers and industry members. Making smart and efficient use of such tools, professionals can connect with other business colleagues, share information, experiences and finally create opportunities for newcomers.

Chapter 5

Knowing drinking water and wastewater systems

The present section focuses on drinking water and wastewater systems. A glossary of common water and wastewater terminologies used in the industry is provided in **Appendix A** for reference purposes. Technical aspects discussed in this chapter include:

- Drinking Water Systems (Intake, Treatment and Distribution)
- Wastewater Systems (Collection, Conveyance and Treatment)
- Sludge Treatment and Disposal Methods

5.1 DRINKING WATER SYSTEMS

Drinking water is one of the basic components for the survival and well-being of civilizations. As a profession, the general scope of a drinking water system ranges from intake sources (such as rivers, lakes, reservoirs, etc.) to its treatment and distribution systems for customers (Alaska DEC, 2013; U.S. EPA, 2004; USFA, 2008). Figure 5.1 depicts the basic components of a drinking water system.

5.1.1 Water sources and intake systems

Water wells, springs, rivers, lakes and reservoirs are some of the common sources of drinking water for communities worldwide. In order to provide high quality drinking water, it is important to select intake water sources as well as their locations carefully. Selected source type affects the quality of influent water, available quantity and degree of treatment required.

Typically, large water treatment plants rely on surface water sources such as rivers, lakes or reservoirs whereas smaller treatment systems use groundwater sources. Wells are extremely important for societies where surface water sources are limited and where people thrive on groundwater sources.

(A) Groundwater Sources

- Wells provide groundwater for direct drinking purposes or for further processing at water treatment plants before consumption. In general, there are three types of water wells (U.S. EPA Private Wells; USGS Groundwater Wells; WHO & IRC, 2003):

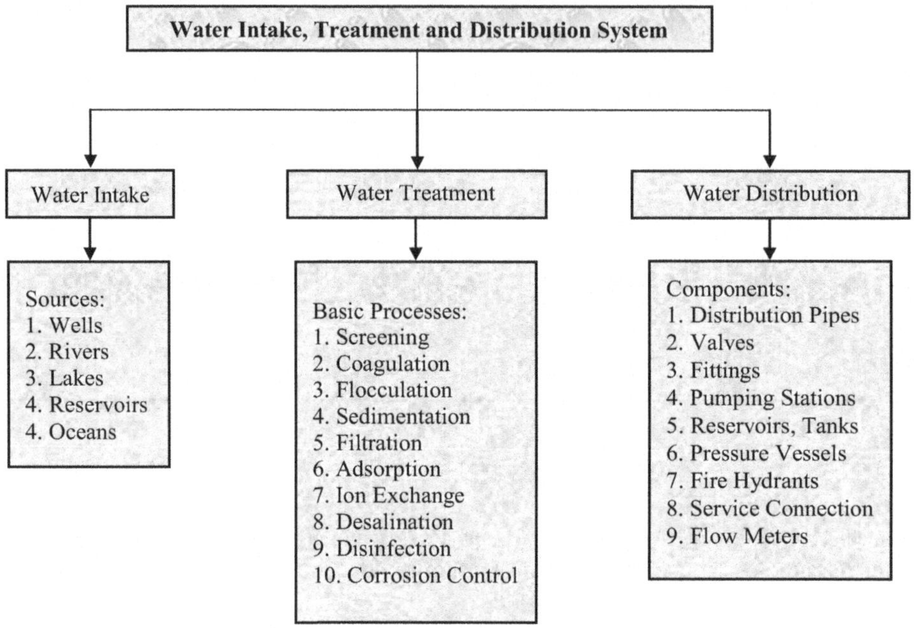

Figure 5.1 Basic components of drinking water system.

- **Dug wells:** When the ground is soft and the water table is shallow, dug wells can be used as a water source. Typically such wells are dug manually with a pick, a shovel and other equipment to a depth below the water table until incoming water exceeds the digger's bailing rate. The main components of a dug well include (i) a stone, brick or concrete apron, (ii) a headwall (the part of the well lining above ground) at a convenient height for collecting water, (iii) a lining that prevents the well from collapsing and (iv) a well cover of wood, stone, or concrete.
- **Driven wells:** Wells of this type are commonly built by driving a small-diameter pipe into soft earth, such as sand or gravel. A screen is usually attached to the bottom of the pipe to filter out sand and other particles. However, such wells can only tap shallow water, and because the source of the water is so close to the surface, contamination from surface pollutants can potentially occur. The well cover is typically a tight-fitting concrete curb and cap with no cracks, and sits about a foot above the ground.
- **Drilled wells:** Modern drilled wells require fairly complicated and expensive drill rigs. Drill rigs use rotary drill bits that crush rocks away, percussion bits that smash the rock, or, larger auger bits if the ground is soft. The main parts of the drilled well typically include (i) a concrete apron around the borehole at ground level to prevent surface water entering the well, (ii) a lining below ground, but not going into the aquifer, to prevent it from collapsing, (iii) a slotted pipe below water level, to allow groundwater to enter the well, surrounded by gravel and (iv) a well cover. Drilled wells can be drilled deeper and often use a pump placed at the bottom to push water up to the surface.

(B) Reservoirs, Lakes, Rivers and Streams

- Common surface water sources include impounding reservoirs, fresh-water lakes, rivers and streams. Reservoirs are created by placing dams across rivers or streams for continuous water supply. Lakes are formed by the collection of water in a natural basin or depression. These lakes are supplied by countless streams and rivers, which, in turn, are supplied from water runoff from hills (USFA, 2008).
- Typical surface water intake structures may include small dams, submerged or floating intakes, bottom intake, protected side intake, spring intake, infiltration galleries, and so on. Primary functions of these intake structures is to provide high water quality from the source (Alaska DEC, 2013; WHO & IRC, 2003).
- Surface water sources from human-inhabited environments generally contain high turbidity and possibly some taste- and odor-producing substances (USFA, 2008). Water from surface sources is often contaminated by microbes, whereas groundwater is normally safer, but even groundwater can be contaminated by harmful chemicals from human activities or from the natural environment (WHO & IRC, 2003).
- In general, surface water is softer as compared to groundwater and hence easier to treat. However, depending upon precipitation, water turbidity can fluctuate significantly. An increase in water turbidity value results in correspondingly increased treatment costs (Alaska DEC, 2013).
- Ambient temperature changes can result in surface water temperature fluctuations and this could affect the treatment process and water quality. Also, in cold weather countries, wintry conditions may cause intake structures to get clogged or damaged due to ice, or the source may be shallow enough to get completely frozen in the winter (Alaska DEC, 2013).

(C) Ocean Intake Sources

- The purpose of desalination plant intakes is to collect source seawater of adequate quantity and quality in a reliable and sustainable fashion so as to produce desalinated water cost-effectively with minimal impact on the environment. Two types of water intakes widely used for desalination plants include open surface intakes and subsurface intakes (Water Reuse Association, 2011).
- **Surface intakes:** Open surface intakes collect seawater directly from the ocean via an on-shore or off-shore inlet structure and the pipeline interconnecting this structure to the desalination plant. Surface intakes are located above the seafloor and are the most common type of intake for large plants. A typical open intake structure includes intake screens, conveyance piping, and a wet well or other mechanism for housing the system pumps. Intake trash racks and screens are provided to respectively remove debris and particulate matter. Surface intakes also require additional pre-treatment due to the presence of marine life and small particles that must be removed before the desalination process (WRF, 2011; Water Reuse Association, 2011).
- **Subsurface intakes:** Subsurface intakes, such as vertical beach wells, horizontal wells, slant wells and infiltration galleries tap into the saline or brackish coastal aquifer and/or the off-shore aquifer under the ocean floor. Subsurface intakes are buried pipes and/or wells buried beneath the beach/ocean floor. Compared to surface intakes, subsurface intakes are typically limited in capacity due to local geology; however, the extensive pretreatment required for surface intakes is either eliminated or greatly reduced because the subsurface acts as a natural filter (WRF, 2011; Water Reuse Association, 2011).

5.1.2 Water treatment processes

The amount and type of treatment required at a public water treatment system depends upon the source type and quality. Three common classifications for water supplies include (adopted from USFA, 2008):

(1) **Fresh water:** Water source is typically surface water or from the natural runoff of water through the water cycle. By definition, these waters are not 'brackish', are not to be considered 'polluted', and have no trace of salt.
(2) **Salt water:** Traditionally, salt water from sea has not been used as a source for domestic consumption. With recent developments in reverse osmosis and filtering technology to remove salt content, ocean waters are looked at as an alternative source.
(3) **Reclaimed water:** Regulations require sewer water to be treated before it can be discharged into lakes, rivers and so on. Treated effluent water can be 'reclaimed' by going through another treatment plant process for re-entry into a municipal water system. The costs are significantly high and scientific work is progressing in this area.

Many water treatment plants use a combination of coagulation, sedimentation, filtration and disinfection processes to provide clean, safe drinking water to the public. A brief description of the typical water treatment processes involved follows (SDWF, Water factsheet; U.S. EPA, 1999a, 2004; USFA, 2008).

- **Screening** – Screening involves removal of larger solids and debris (such as twigs, leaves, paper, stones and other foreign matter) from influent water and assisting with turbidity reduction to a certain extent. Screens are frequently backwashed to prevent clogging (USFA, 2008).
- **Pre-sedimentation** – Pre-sedimentation is commonly used when substantial amounts of sand and gravel may be present in the source water or when raw water turbidity is high. Depending on the purpose, pre-sedimentation basins may be relatively large settling ponds or small concrete basins. While the water moves slowly through reservoir, much of the sand and silt settles to the bottom. Chemical addition is often used to enhance solids removal. Treatment lines and basins are shut down periodically during times of minimum domestic consumption for cleaning (U.S. EPA, 1999).
- **Coagulation/Rapid Mixing** – The coagulation process involves conditioning of suspended solids particles to promote their agglomeration and produce larger particles that can be removed by consolidation and settling in subsequent processes. Coagulant (such as aluminum sulfate, ferric sulfate, ferric chloride or polymer), is added to the water as it flows to sedimentation basins. Following coagulation, the processes of flocculation, sedimentation and filtration are used to remove 'destabilized' particles from suspension (SDWF, Water factsheet; U.S. EPA, 1999).
- **Flocculation** – Flocculation refers to water treatment processes that combine or coagulate small particles into larger particles, which settle out of the water as sediment. Flocculation is commonly in conjunction with the coagulation process. Flocculation is generally accomplished by mixing destabilized suspension to provide the opportunity for the particles to come into contact with one another and stick together as larger 'floc' particles. The water is gently stirred with large paddles to distribute the coagulant (U.S. EPA, 1999, 2004).
- **Sedimentation** – This process involves removal of solids from water by means of gravity separation. After coagulation/flocculation processes, water flows into sedimentation basins where relatively quiescent conditions prevail and particles settle to the bottom. Clear water passes out of the basin over an effluent baffle or weir for further treatment (U.S. EPA, 1999, 2004).
- **Filtration** – During the filtration process, water flows through the filter media bed and solids get physically retained on the media. This unit operation helps to remove smaller particles and reduce water turbidity substantially. Filters are backwashed to periodically remove solids collected on

the media. Particles removed typically include clays and silts, natural organic matter, precipitates from other treatment processes in the facility, iron and manganese, and microorganisms. Filtration clarifies water and enhances the effectiveness of disinfection (U.S. EPA, 1999, 2004).

- **Membrane Processes** – Membrane units are similar to filters except that a membrane is used instead of the filter media. During treatment, water is passed through a semi-permeable membrane for the removal of solids. In addition, membrane processes also remove dissolved organic or inorganic constituents. Common types of membrane processes include microfiltration, ultrafiltration, nanofiltration, and reverse osmosis (SDWF, Water factsheet; U.S. EPA, 1999).
- **Absorption** – The adsorption process uses granular or powder activated carbon for removing organic contaminants, unwanted coloring, and taste-and-odor-causing compounds from the drinking water (U.S. EPA, 2004).
- **Ion Exchange** – Ion exchange treatment processes are used for removing inorganic contaminants when they cannot be removed adequately by filtration or sedimentation processes. Ion exchange systems are also used for treating hard water as well as for removal of chemicals such as excess fluorides, arsenic, chromium, nitrates, uranium or radium (U.S. EPA, 2004).
- **Disinfection** – Water is often disinfected before entering the distribution system, to ensure that potentially dangerous microorganisms are killed. Chlorine, chloramines, and chlorine dioxide are the common disinfectants used. For relatively clean water sources, ultraviolet radiation and ozone disinfection are alternative treatment methods, however neither of these techniques can effectively control biological contaminants in the distribution pipes. A major challenge for water suppliers is balancing the risks from microbial pathogens and generation of disinfection byproducts (U.S. EPA, 2004).
- **Additives** – Depending on the treated drinking water quality, additives may be injected into the water streams to accomplish additional benefits. For example, fluoride addition to reduce tooth decay and calcium hydroxide addition to reduce corrosion in the pipes and the equipment of the distribution system (USFA, 2008).

Depending upon water source characteristics, the requirements for drinking water quality and operation complexity, additional treatment processes or chemicals could be involved at the treatment plant. Appendix H includes a list of the various components of the water processing units for reference.

5.1.3 Water distribution system

The distribution system conveys water from a centralized treatment facility to the customers that is, homes, schools, hospitals, businesses, restaurants, industries, and so on, in the community. Distribution network designs also ensure that water is delivered at the desired working pressure to the consumers and provisions are made for fire water required under emergency situations.

Smaller water distribution systems serving a handful of households may be relatively simpler, whereas water systems for large cities or a metropolitan area can be extremely complex in nature, sometimes involving thousands of miles of piping serving millions of people (U.S. EPA, 2004).

Distribution systems typically consist of an interconnected piping network, valves, fittings, storage tanks or reservoirs, fire hydrants, pumping systems, booster stations, service connections, and so on. The water distribution network should be able to furnish instantaneous flow demands at all times required under all flow conditions. Consumer demands and fire suppression needs determine water distribution pipe sizes, pumping station capacities, and storage tanks or reservoir volumes (Alaska DEC, 2013; USFA, 2008).

Even though high quality of water is produced at the water treatment plant, it is important to make sure that water does not get contaminated within the distribution system owing to factors such as pressure problems, microorganism growth, water main breaks, and so on. (U.S. EPA, 2004). A brief description of basic components involved in a water distribution system follows:

- **Distribution Network:** Distribution network piping can vary from simple structures to extremely complicated systems. Water mains can be laid out in loops, grids or branch format. A looped network system results in overall smaller pipe sizes and lower water usage than either the regular grid system or the branching network system. However, a branching network results in comparatively lower costs than the other networks because of reduction in total pipe length, despite larger pipe diameters. Grid or loop layouts provide greater fire suppression flows and result in fewer dead-end lines. Whereas the branching network system results in a number of 'dead-end lines' potentially leading to taste, odor or bacteriological problems (Alaska DEC, 2013; USFA, 2008).
- **Water Piping:** Pipes used for drinking water distribution need to be suitable for the transportation of water. In many countries, standards have been established for the pipe materials and construction methods used. When in contact with water or soil, the material should be corrosion resistant to possible chemical reactions. Furthermore, the pipes should be able to handle specified internal and external pressures. Different types of materials commonly used for water pipes, distribution mains, home plumbing, reservoirs and tank connections, and so on, are as listed below (Alaska DEC, 2013; Tomboulian *et al.* 2004):
  (1) Asbestos Cement
  (2) Cast Iron
  (3) Chlorinated Polyvinyl Chloride (CPVC)
  (4) Copper
  (5) Ductile Iron
  (6) Galvanized Steel
  (7) Glass Fiber Reinforced Plastic (GRP)
  (8) High Density Polyethylene (HDPE)
  (9) Lead
  (10) Polybutadiene
  (11) Polyvinyl Chloride (PVC)
  (12) Reinforced Concrete
  (13) Steel
  (14) Wood

  The selection of pipe materials may vary from region to region and also depend upon availability of sources, economic conditions, funds allocated for water infrastructures and other related factors.
- **Pipe Fittings:** Pipe fittings are commonly used in the network for changing sizes or flow directions of waterlines or connecting other accessories. Typical fittings used for water systems include reducer, increaser, wye, tee, bend, cross, adapter, and so on. Depending upon material type, the connectors can be flanged, mechanical joint, hub, threaded, butt, socket welded or glued (Alaska DEC, 2013).
- **Valves:** Valves are used to control, regulate or isolate water flows within the distribution network system. Various types of valves commonly used for the water system include (Alaska DEC, 2013; USFA, 2008):
  (1) Check valve
  (2) Butterfly valve
  (3) Gate valve
  (4) Plug valve

(5) Ball valve
(6) Globe valve
(7) Air Relief valve
(8) Air and vacuum valve

- **Water Storage Tanks:** Common materials for storage tanks or reservoirs include steel, concrete, wood, plastic and fiberglass. Finished water storage reservoirs are generally used in water distribution system and serve several purposes, as listed below (Alaska DEC, 2013):
  (1) Elevation of tanks or reservoirs help to maintain pressure on the system
  (2) Provides flow during peak demand
  (3) Provides fire water demand
  (4) Provides surge relief associated with pumping start-stop cycles
  (5) Provides or increases detention time
  Typical types of reservoir:
  (1) Below ground reservoir
  (2) At grade reservoir
  (3) Elevated tanks or standpipes
  (4) Hydropneumatic tanks (air pressurized)
- **Fire Hydrants:** Fire hydrants provide water for fire suppression events and mainly are of two types, dry barrel and wet-barrel. These hydrants can serve other purposes as well, such as providing water for street cleaning, construction projects, sewer cleaning or even testing distribution system's pressure and flow capacities in the area.
- **Water Meters:** Meters help to monitor flows within the distribution network and at the same time assist to fairly distribute water service costs based on water consumption by the customers. In addition, meters can help operators to understand daily water consumption rates and identify any unaccounted water in the system and losses due to leakages (Alaska DEC, 2013).
- **Ancillaries Equipment, Linings and Coatings:** Other ancillary equipment includes gaskets, o-rings, sealants, lubricants, paints, primers, additives, adhesives, solders, and so on. Linings and coatings for water systems may be epoxy, polyester, polyurethane, polyacrylate resins, cement mortar, bituminous materials, and so on, (Tomboulian *et al.* 2004).

Most of the water distribution components are required to follow specifications and standards related to the material quality, pressure ratings, installation methods and other relevant parameters as recommended by the regulatory authorities.

5.2 WASTEWATER SYSTEMS

Population growth, increased water usage followed by scarcity of new water sources and need for improved sanitation have resulted in significant attention towards wastewater treatment and its reuse. From the profession point of view, typical wastewater system consists of the following components (Libabher & Orozco-Jaramillo, 2012; Lombardo, 2004; OR DEQ; U.S. EPA, 2000a–e, 2002a–c, 2003a,b, 2005a, 2006b, 2013; van Haandel & van der Lubbe, 2012):

(1) Sewer collection and conveyance systems
(2) Industrial pretreatment systems
(3) Onsite, clustered and centralized treatment systems
(4) Sludge treatment and disposal systems

5.2.1 Wastewater collection and conveyance systems

Sewer collection and conveyance system components commonly used by communities include gravity sewers, pump stations, pressure sewers and vacuum sewers or a combination therein (Figure 5.2).

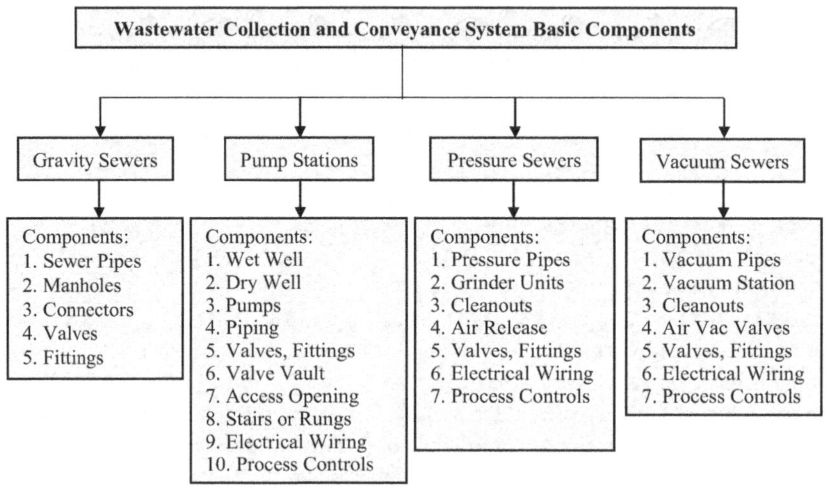

Figure 5.2 Basic components of wastewater collection and conveyance systems.

The selection and design of a specific wastewater collection system depends on several factors such as flows, industrial discharges, inflow and infiltration, project size, topography, sensitive natural resources, depth to bedrock or groundwater, distance to the treatment and dispersal site, housing density, road frontage, project costs, environmental impacts, and so on (Ohio EPA, 2013; Lombardo, 2004).

(A) Gravity Sewer System

- A gravity sewer system takes advantage of the elevation difference between wastewater generation sources and the destination wastewater treatment system based on the topography and grading of the service area. Hydraulic heads available assist in conveying wastewater by gravity via sloped pipes.
- Important parameters considered for the design of gravity sewer systems may include design average and peak flow requirements, topography and available slopes in the area, site conditions, excavation needs, pipe depths, pipe sizes, velocities, serviceability, and so on (U.S. EPA, 2002a).
- Manholes are typically installed within the gravity collection system for connecting pipes from different areas and changing flow directions. These manholes can also serve as sampling locations, to analyze flow conditions, water quality, leakage issues, and so on (Bertrand-Krajewski *et al.* 2006; Geller *et al.* 2007; U.S. EPA, 2002a).

(B) Pump Stations

- Pump stations (PS) are usually provided in the gravity sewer system when sufficient head is not available to convey wastewater from the collection area due to the existing grading. In such cases, a pump station is provided to elevate the water head and transfer it to a higher location in the sewer collection system. Pumping and lift stations can substantially increase the cost of the collection system (U.S. EPA, 2002a).
- Some of the common types of pump stations include the dry well pump station (PS), submersible PS and suction lift PS.

- ◦ **Dry-Pit or Dry-Well Pump Station:** In this configuration, pumps and associated assembly are installed in a dry pit or dry well (i.e., pump room). Whereas, the wet-well receiving wastewater is a separate chamber attached or located adjacent to the dry-well. When a dry well is located below ground level, mechanical ventilation is required. The main advantage of a dry-well pump station is the easy access for routine maintenance and visual inspection of the system as compared to the submersible pump stations (Ohio EPA, 2013; U.S. EPA, 2000b).
- ◦ **Submersible Pump Station:** This type of pump station does not have a separate pump room and the pumps are submerged under the receiving wastewater in the wet well. Pump station header piping, associated valves and flow meters are located in a separate vault at grade for easy access. For maintenance and access, the submersible pumps are periodically removed and reinstalled with the help of guide rails and hoists. Hence, the pump and motor units are removed and installed from above ground without dewatering or having an operator enter the wet well. Submersible pump stations normally do not have large above-ground structures and hence blend in with their surrounding environment in residential areas (Ohio EPA, 2013; U.S. EPA, 2000b).
- ◦ **Suction Lift Pump Stations:** Suction lift pump stations use a self-priming type of pump. The pump equipment compartment is located above grade or offset and is effectively isolated from the wet well to prevent any hazardous and corrosive sewer atmosphere from entering the equipment chamber (Ohio EPA, 2013).

- Efficient wastewater pump station designs should select pumps suitable for all design and worst case flow conditions, keep pipework layout short, simple and self-venting, arrange for back flushing and provide access and means for removing scum, sediment and debris (Brandt *et al.* 2011).
- In general, pump stations should also include pump controls, security fencing, access hatches, visual and audible alarms (such as for water levels, pump failures, loss of pump), backup power sources, and so on (Ohio EPA, 2013).

(C) Pressure Sewers (U.S. EPA, 2002c)

- Low-pressure sewer systems are useful for specific conditions in which conventional gravity sewer systems are not suitable and costly.
- Two major types of pressure sewer system include a septic tank effluent pump (STEP) system and grinder pump (GP) system. In the STEP system, wastewater flows into a conventional septic tank for capturing solids. Liquid effluent then flows into a holding tank containing pump and control devices. Effluent is then pumped and transferred for treatment. Whereas in a GP system, sewage flows to a vault where a grinder pump grinds solids and discharges sewage into a pressurized pipe system.
- Since pressure sewer pipe sizes and depth requirements are comparatively less, material and trenching costs are significantly lower. Pressure sewers reduce infiltration into the system. Also, since wastewater is transported under pressure, more flexibility is provided for the location of the treatment plant.
- Operation and maintenance (O&M) costs for pressure sewers are often higher than those for conventional gravity systems. However, lift stations in a conventional gravity sewer can reverse this situation.

(D) Vacuum Sewers (WERF, 2009)

- A vacuum sewer system uses a vacuum (negative pressure) to collect wastewater from multiple sources within the sewer system and convey it to a central location for treatment. This differential pressure is developed by a central vacuum station.

- Typically, sewage from one or more homes or businesses flows by gravity into a small valve pit. A service line connects the valve pit to the main vacuum line. Each valve pit is fitted with a pneumatic pressure-controlled vacuum valve. This valve automatically opens after a predetermined volume of sewage has entered the sump. The difference in pressure between the valve pit (at atmospheric pressure) and the main vacuum line (under negative pressure) pulls wastewater and air through the service line.
- Vacuum stations may include two or more vacuum pumps and a large vacuum tank. The tank at the vacuum station holds the vacuum on the collection network and prevents the vacuum pumps from having to operate continuously.
- When sewage reaches the vacuum station, it flows into a collection tank. Sewage pumps are then used to convey the collected sewage through a force main to the treatment component. Like gravity sewers, vacuum sewers are commonly installed on a slope toward the vacuum station.

(E) Pipe Materials

Common pipe materials used for wastewater collection and conveyance systems include (U.S. EPA, 2000d):

(1) Ductile iron pipe (DIP)
(2) Concrete pipe (pre-stressed and reinforced)
(3) Polyvinyl chloride (PVC) pipe
(4) Polyethylene (PE) pipe
(5) High density polyethylene (HDPE) pipe
(6) Polybutylene (PB) pipe
(7) Acrylonitrile-butadiene-styrene (ABS) pipe
(8) Fiberglass reinforced plastic (FRP) pipe
(9) Vitrified clay pipe

- Various sewer pipe fittings include tee, bend, elbow, wye, reducers, increasers, cap, coupling, adapter, plug, and so on. Normally, these components need to follow specifications and standards set by authorities related to the material quality, pressure ratings, installation methods and other relevant parameters.

5.2.2 Industrial pretreatment programs

- Community treatment plants typically collect wastewater from homes, commercial buildings, offices, restaurants, and other domestic sources. Generally, these wastewater treatment plants (WWTPs) are designed to treat this domestic wastewater.
- The wastewater characteristics of non-domestic sources such as industries and certain commercial entities could be significantly different from that of domestic wastewater and may contain potential toxic or non-conventional pollutants that affect performance of WWTPs if not-pretreated before discharging into the sewer system.
- Federal, state and local authorities establish regulations for pre-treatment of wastewaters from these peculiar industrial or commercial sources. The basic objectives of the industrial wastewater pretreatment program may include (adopted from: OR DEQ, Industrial Pretreatment Program):
    - Protecting WWTP from pollutants that may interfere with plant operations.
    - Prevent pollutants into WWTP that can cause pass through of untreated pollutants to receiving waters.
    - Manage pollutants entering WWTP to improve opportunities for reuse of treated wastewater and residuals (sewage sludge).

- Prevent pollutants to WWTP that can cause worker health or safety concerns, or pose potential endangerment to the public or to the environment.

5.2.3 Onsite, clustered and centralized wastewater treatment systems

Three basic alternatives for the treatment of wastewater include (Lombardo, 2004; U.S. EPA, 2002b; U.S. EPA Handbook, 2005a):

- Onsite wastewater treatment systems
- Clustered wastewater treatment systems
- Centralized wastewater treatment systems

Onsite systems are common for low-density communities, rural areas and small commercial developments and typically include collection, treatment and disposal of treated effluent on an individual's property (Lombardo, 2004).

Clustered wastewater treatment systems serve selected wastewater connections in specific areas. Smaller cluster systems resemble somewhat the onsite systems whereas larger cluster systems serving bigger communities tend towards centralized plants (Lombardo, 2004).

Centralized treatment systems are associated with high-density communities and developments such as cities and commercial areas. These facilities receive and treat wastewater from a community's common collection and conveyance systems. Treated effluent is discharged to surface water or groundwater sources (Lombardo, 2004). Figure 5.3 provides an overview of wastewater treatment system components including onsite, clustered and centralized systems.

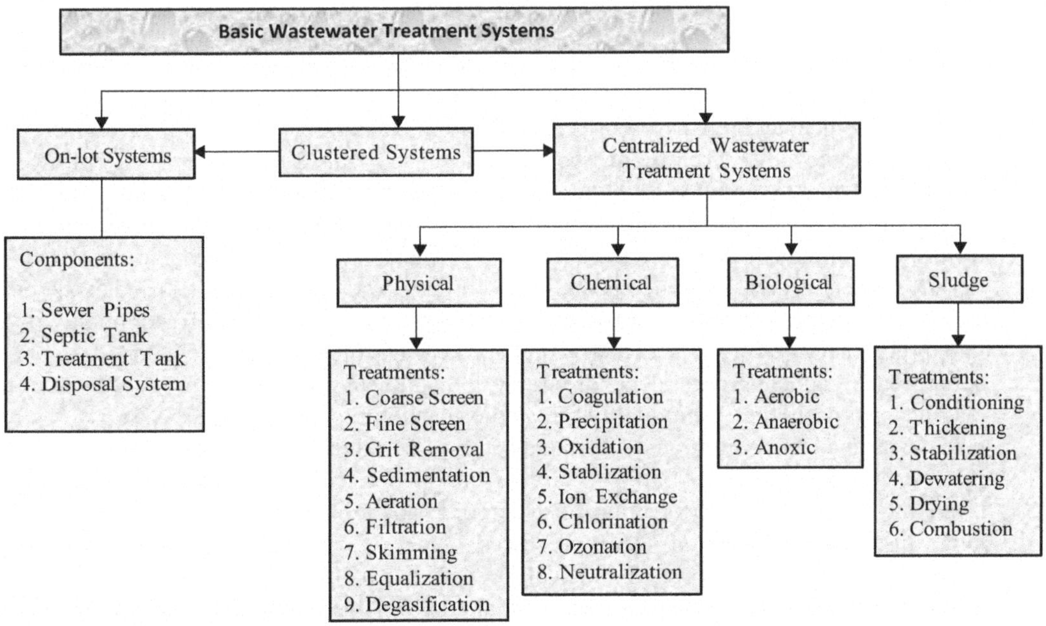

Figure 5.3 Basic components of wastewater treatment systems.

(A) On-Site Treatment Systems (U.S. EPA, 2002b)

- Onsite treatment systems primarily consist of a septic tank and a soil absorption field, also known as a subsurface wastewater infiltration system. Sometimes, they can also include advanced treatment systems and drip dispersal systems.
- Septic tanks remove most settleable solids, oils, greases, and floatable material and function as a bioreactor that promotes partial digestion of retained organic matter. Septic tank effluent has traditionally been discharged to soil, sand, or other media absorption fields for further treatment through biological processes, adsorption, filtration, and infiltration in the underlying soils.
- These systems work well if: installed in areas with suitable soils and hydraulic capacities; are designed to treat the incoming waste load to meet public health, groundwater, and surface water performance standards; are installed properly; and maintained to ensure long-term performance.

(B) Clustered Wastewater Treatment Systems

- Clustered systems are typically pursued when onsite treatment systems are not technically feasible and sewer connections to centralized plant are expensive, unavailable, or not desired due to unfavorable environmental impacts (Lombardo, 2004).
- Cluster systems enable communities to develop a wide array of wastewater solutions that can optimize economic and environmental objectives and avoid difficulties and adverse impacts of solely centralized or onsite options (Lombardo, 2004).
- The selection of appropriate technologies depends upon factors such as influent wastewater characteristics, effluent requirements, capital and O&M costs, flow variations, system reliability, footprint needs, location requirements, and so on (Lombardo, 2004).
- Treatment options for cluster plants can include various methods available for onsite and centralized systems. Systems may involve simpler, passive treatments with soil dispersal (septic systems), or more complex and mechanized advanced fixed film or suspended growth biological processes (Lombardo, 2004; EPA Handbook, 2005a).
- The benefits of clustered systems typically include sewage treatment often at lower cost as compared to centralized plants, water conservation by recharging groundwater aquifers and maintaining dry season flow in streams, flexible wastewater treatment options and comparatively lower replacement and repair costs (EPA Handbook, 2005a).

(C) Centralized Wastewater Treatment Systems (Libabher *et al.* 2012; Ohio EPA, 2013; SDWF, Wastewater factsheet; U.S. EPA, 1999a–c, 2000a–e; van Haandel & van der Lubbe, 2012)

- A centralized wastewater treatment plant serves a larger population and typically receives sewage from the community's sewer collection and conveyance system. The generation sources may include residential areas, commercial buildings, offices, restaurants, industries and other sources.
- The facility involves extensive wastewater treatment processes, which are commonly divided into preliminary, primary, secondary and tertiary treatments. Selection and design of an appropriate treatment process depends largely upon influent wastewater flow rates, contaminant loadings and desired effluent quality. Treated effluent is typically discharged to river, stream, lake, so on and can be infiltrated to groundwater or reused for irrigation purposes.
- Preliminary and primary treatments are based on physical processes. These may include screening or grit systems (for removal of large, floatable materials and rags), sedimentation tanks, dissolved air floatation (DAF) processes.
- Secondary treatment involves biological processes carried out commonly by aerobic or anaerobic bacteria for the removal of colloidal and dissolved organic matter (BOD). The two main types of biological systems are suspended growth reactors and attached growth reactors.

- Tertiary processes are advanced treatment methods that typically include nitrification and denitrification for nitrogen removal and combination of biological processes and chemical precipitation for phosphorus removal from the wastewater. Processes can also remove color, metals and organic chemicals.
- The disinfection process removes pathogens and other bacteria from treated effluent. Disinfection methods include chlorination, ozonation, ultra-violet (UV) radiation and similar systems.
- *Appendix I* comprises a list of the common components of a wastewater treatment process.

(D) Wastewater Treatment Technologies

Several advancements related to wastewater treatment technologies are going on in different parts of the world. In general, these technological advancements can be classified as at research and development stage, emerging, innovative and established. Following is a partial list of various wastewater biological treatment technologies practiced in the field and referenced from '*Emerging Technologies for Wastewater Treatment and In-Plant Wet-Weather Management, 2013*' publication by United States Environmental Protection Agency's Division of Water (U.S. EPA, 2013). For further details, please refer to this document:

(1) Anaerobic Attached Growth Systems
(2) Anaerobic Contact Processes
(3) Anaerobic/Anoxic/Oxic (A2/O)
(4) Bardenpho® (Four Stage)
(5) Bardenpho® (Five Stage)
(6) Biological Aerated Filters (BAF)
(7) Contact Stabilization Process
(8) Conventional Extended Aeration System
(9) Denitrification Filter
(10) Facultative and Aerated Lagoons
(11) Fluidized Bed Bioreactors (FBBR)
(12) Integrated Fixed-Film Activated Sludge (IFAS)
(13) Intermittent Cycle Extended Aeration System
(14) Johannesburg Process
(15) Ludzack-Ettinger Process
(16) Membrane Bioreactor (MBR)
(17) Modified Ludzack-Ettinger (MLE) Process
(18) Moving Bed Bio-Reactor (MBBR) Process
(19) Orbal™ Process
(20) Oxidation ditches/Aerated lagoons
(21) Phoredox (Anaerobic/Oxic [A/O])
(22) Phostrip Process
(23) Rotating Biological Contactors (RBCs)
(24) Schreiber™ Process
(25) Sequencing Batch Reactor (SBR)
(26) Staged Activated Sludge Process
(27) Submerged Rotating Biological Contactors (SRBC)
(28) Trickling Filter (TF)
(29) Trickling Filter/Solids Contactor (TF/SC)
(30) SNdN Process*
(31) Step Feed (Alternating Anoxic and Aerobic)
(32) Step Feed Biological Nutrients Removal (BNR)
(33) Upflow Anaerobic Sludge Blanket (UASB)
(34) Wuhrman Process

* – SNdN – Simultaneous Nitrification denitrification.

5.3 SLUDGE TREATMENT AND DISPOSAL METHODS

- 'Sewage Sludge' includes any solid, semisolid, or liquid residue removed during the treatment of municipal wastewater or domestic sewage. Sewage sludge includes solids removed during primary, secondary, or advanced wastewater treatment, scum, septage, portable toilet pumpings, and so on (Ohio EPA, 2013). 'Biosolids' specifically refers to sewage sludge that has undergone treatment and stabilization to meet government standards for its beneficial use (U.S. EPA, 2003b).

- Sludge processing may include conditioning, thickening, stabilization, dewatering, thermal conversion and drying. 'Emerging Technologies for Biosolids Management' report by United States Environmental Protection Agency, Office of Water, 2006, discusses various established, innovative and embryonic technologies in the sludge treatment area. Following is a partial list of methods practiced in the industry for sludge processing (EU, 1999; U.S. EPA, 2002b, 2003a,b, 2006b):

No.	Treatment method	Technologies
1	Sludge conditioning	• Chemical • Thermal
2	Sludge thickening	• Centrifuge • Flotation thickening • Gravity thickening • Gravity belt thickening • Rotary drum thickening
3	Sludge stabilization	• Aerobic digestion • Alkaline stabilization • Anaerobic digestion • Composting • Pasteurization • Solidification
4	Sludge dewatering	• Belt filter press • Centrifuge • Chamber press • Drying beds • Vacuum filters
5	Sludge thermal conversion	• Combustion • Oxidation • Pyrolysis/gasification • Vitrification
6	Sludge drying	• Direct drying • Flash drying • Indirect drying

- Stabilized sludge classification is usually based on degree of treatment and stabilization achieved. Classification and treatment methods can vary for states, regions and countries. Stabilized sludge after treatment is commonly used as fertilizer or soil conditioner, disposed of in a landfill, or incinerated in a sewage sludge incinerator with the ash disposed of in a landfill. Here are some common use and disposal options for stabilized sludge (EU, 1999; U.S. EPA, 2003b; U.S. EPA, 2002b; U.S. EPA Emerging Biosolids Technologies, 2006b):

  Land Based Sludge Use Options
  - Agriculture
  - Reclamation
  - Lawns
  - Gardens

- Forestry
- Landscaping
- Silviculture
- Horticulture

Fuel Based Sludge Use Options

- Incineration
- Supplementary Fuel
- Gasification

Sludge Disposal Option

- Landfill

Chapter 6

Understanding drinking water and wastewater projects

Now that we have reviewed professional progression, drinking water and wastewater industry scope, types of systems, and treatment technologies, the next important step is to understand how projects are executed. Although projects can be of varied types depending upon specific profession and job profile, the present chapter mainly focuses on the drinking water and wastewater infrastructures projects. In a broader sense, this chapter discusses client types, projects and execution phases involved followed by system operations (Figure 6.1).

6.1 CLIENTS OVERVIEW

In general, clients in the drinking water and wastewater industry can be categorized based on criteria such as institution type, funding source (private or government), geographical presence (local, national or global) and so on.

Organizations served: Drinking water or wastewater industry clients may include research organizations, academic institutions, municipal authorities, land developers, manufacturers, industries, international establishments, regulatory authorities, and so on. Each client handles peculiar projects and requires specific professional services.

Private or Government: Just like the type of industry served, it is also imperative to get familiar with private and government sector clients as they have different set-ups, modes of functioning and budgetary needs.

Geographical presence: Many of the drinking water and wastewater projects requiring professional services are handled by local governments or industries. However, there are several clients requiring large-scale or global level technology, design or policy related project services that need international work experience and vision. Accepting such assignments by services firm depends upon availability of knowledgeble staff, their experience, expertise and exposure to wider industry know-how.

Type of services pursued: Services pursued by clients may include planning, research and development, laboratory testing and analysis, pilot-scale studies, sales, business expansion, and so on (Burke *et al.* 2013; Park *et al.* 2012; Rousseau & Tranchant, 2002; Suzuki & Minami, 1991) as well as preparation of feasibility study, preliminary design, final design, opinion of cost, bidding, construction administration, construction observation, environmental review, grant writing, funding application development, project administration, permit application, surveying, and so on (Lamont & Young, 2007).

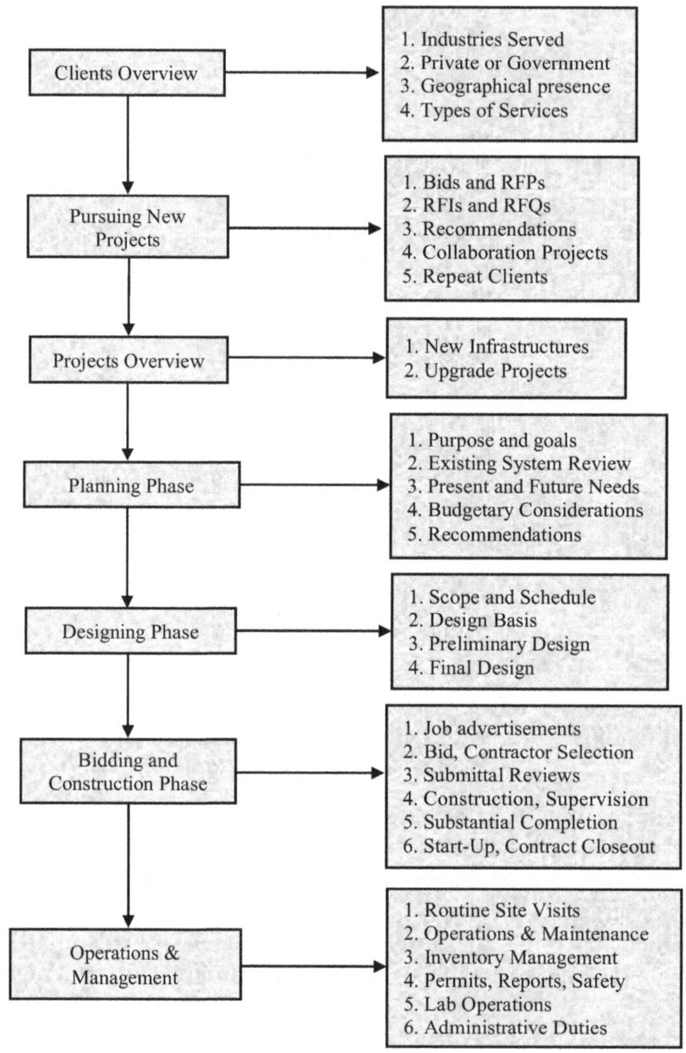

Figure 6.1 Overview of water and wastewater infrastructure projects.

Projects by townships, boroughs, municipalities and other government entities are typically funded via tax-payer's money and hence get social, political, financial and environmental attention within the community. Private organizations such as utility companies, developers, research institutes, industries or manufacturers may also look for specific services in this sector. Successful professionals or consultants understand client needs very well and can develop good long-term relationships with successful and efficient execution of projects.

6.2 PURSUING NEW PROJECTS

From the client's point of view, the quality of services obtained has a considerable impact on the ultimate efficiency, economy and effectiveness of the projects. Hence, the individual or services firm must be chosen based on accountability, reliability, responsibility, skill, education and training, judgment, integrity, and moral worth (Lamont & Young, 2007). Service procurement processes may vary significantly for different organizations, regions and countries.

One the other hand, consultants or services firms may pursue new drinking water and wastewater projects using various approaches, such as by responding to solicitations advertised by clients, through recommendation by previous clients, collaborating with other firms or via continued client relationships and are briefly discussed below.

RFP and bids: Request for Proposal (RFP) or Invitation for Bid are common methods practiced in the drinking water and wastewater industry where solicitations are advertised in newspapers, or other forms of media and proposals or quotations are requested from the contractors or services firms for proposed projects. Documents requested by client may include original bids, proposals, specifications, technical capabilities, insurance information, bonds and all other relevant data. The sealed bids are then opened by the officials, usually in the presence of one or more witnesses, at the time and place specified in the solicitation. The amount of each bid gets recorded and bids are made available for public. Such procedures help clients to obtain the best services on a competitive basis (Commonwealth of Massachusetts, 2006; Lamont & Young, 2007; CIPS & NIGP, WSDOT, 2007).

RFI or RFQ: Many times prior to requesting proposals or bids, clients prefer to request various pieces of information or statement of qualifications from potential product or service providers to compare and review their capabilities for specific projects. These Request for Information (RFI) or Request for Qualifications (RFQ) processes help clients to understand qualifications of probable teams or consultants and can gauge potential competition in the marketplace, prior to issuing the solicitation (Commonwealth of Massachusetts, 2006; Lamont & Young *et al.* 2007; CIPS & NIGP; WSDOT, 2007).

Repeat clients: Quite often, service companies rely on their already established relations with clients for receiving continued work and projects. With good relations, clients may look at service companies as the obvious choice and in such cases avoid advertising competitive solicitations for the new or upcoming projects.

Collaborative projects: Depending upon project size and scope as well as company's capabilities, expertise and limitations, projects can be pursued and executed by collaborative efforts between two or more services firms (Arthur Scott *et al.* 1996; Bassan *et al.* 2013; Fattal *et al.* 1989).

Recommendations: Based on the quality of services provided and satisfaction of previous clients, a company's name may get recommended to other potential clients looking for similar water or wastewater services. This is an excellent opportunity for service providers or manufacturers to get connected with new clients, discuss services offered, past achievements, and win confidence for the new projects.

6.3 PROJECTS OVERVIEW

Realizing staff strength, the available skill set, corporate goals and finances as well as liabilities involved, companies judge their capabilities and interest in pursuing specific projects. Following is a brief overview of infrastructure projects commonly encoutered by drinking water or wastewater professionals.

6.3.1 New infrastructure projects

New drinking water or wastewater projects are commonly a result of recent or upcoming community developments, increased infrastructure demands, changes in government policies, need for advanced systems, compliance with updated or new regulations or permits, and so on (Heard *et al.* 2002; van Nieuwenhuijzen *et al.* 2009; Yang *et al.* 2013).

Planning and design-wise, newer projects are generally easier to approach as compared to upgrade projects, although the complexity of projects can vary significantly. New ventures provide more flexibility in terms of site selection, footprint availability, technology selection, equipment installations, traffic regulations, use of advanced and recent knowledge in the field.

6.3.2 Upgrades projects

Comparatively, upgrade projects are more challenging and are a result of ageing utilities or equipment, expansions within the existing community, connection of new developments to existing systems, increased demand for flow capacities and loadings, connection of newer industrial sources, changes in discharge effluent limits, timely repairs or maintenance needs, efficiency improvements, and so on. Upgrades can be for the complete structure or partially for specific sections, locations or equipment within the drinking water or wastewater systems (Ambulkar *et al.* 2011; Bell & Abel, 2011; Clark *et al.* 1999; Hahn *et al.* 1990; Karpf *et al.* 2011).

With many of the upgrades planned on existing sites, these projects may encounter complexities such as footprint limitations, presence of other existing structures and utilities in the close proximity, need for unaffected operations of surrounding systems, compliance with water quality and existing permit requirements during project execution, significant coordination with authorities and other utility companies, and so on.

6.4 PLANNING PHASE

Planning is the first step towards project execution. The comprehensive planning process integrates technical, environmental, regulatory, economic, financial, political and management issues related with projects (Lombardo, 2004). Good vision and planning can have significant impact on successful implementation, efficient budget management, improved sanitation, excellent water quality and efficient supply systems. Hence, organizations should devote sufficient time, money and efforts towards developing futuristic plans for the drinking water and wastewater infrastructure systems.

Development of a community or project profile, or statement of existing conditions helps to familiarize stakeholders within the community with past planning, and the need for the water or wastewater management plan. The project profile also helps to identify key issues and problems that will require special focus and determine the availability of key data required for subsequent steps in the planning process (Lombardo, 2004).

Regulatory agencies should also be involved from the outset with the planning process. This includes keeping regulators informed of progress during the planning process to insure that they support and approve of the plan that is being developed. Soliciting recommendations and comments from regulators helps to insure that the planning process achieves a viable plan that can be implemented without requiring any time-consuming and costly revisions (Lombardo, 2004).

6.4.1 Existing systems review

Review of the existing infrastructure is a major task for the planning process. Findings from the review process lead to future utility replacement, upgrades or new projects and hence have a major impact on the project economics. Common activities associated with the review process are discussed briefly below:

Literature review: To accomplish planning objectives and develop a project profile, planners need to compile, synthesize, and present data from a wide variety of sources on many different topics. Information may include flows and demands, population and demographic data, industries, economics, land use, natural and hydrological features, environmental sensitive areas, existing infrastructures, management, institutional, regulatory systems, and so on (Lombardo, 2004; US EPA, 2005a). Literature for existing systems such as technical reports, testing data, analytical results, past projects, inspection sheets, system violations information, previous studies, and so on, provide a sound database for the assessment process. Lack or omission of data would be a deterrent for taking appropriate decisions.

Hydraulic capacity analysis: Be it the collection or distribution systems (such as pipes, conduits, intake structures, flow meters, pump stations, booster stations, force mains, valves, hydrants, meters, fittings, etc.) or water and wastewater treatment processes, these infrastructure items need to be evaluated thoroughly for available hydraulic capacities to ensure that they are able to handle present and future flows. Any issues encountered with present operations should be flagged and thoroughly discussed during the review process (Banasiak, 2008; Brocard *et al.* 2009; Pollert *et al.* 2005).

System performance: Similar to hydraulic capacity analysis, unit processes and operations need to be examined for system performance. This review can be in terms of degree of treatment achieved, process operations, efficiencies, process controls involved and final water or effluent quality. For example, water equipment (coagulation and sedimentation tanks, filtration system, disinfection unit, clear well tanks, booster system, etc.), wastewater equipment (septic tanks, screens, de-gritting units, sedimentation process, biological reactors, sludge digesters, disinfection, aeration, etc.) should be reviewed for their historical operation and performance. Any design related problems, malfunctioning, under performance and other process related issues should be identified during the planning process (Harremös *et al.* 2002; Oliveira & von Sperling, 2011; Rochmadi *et al.* 2010).

Structural analysis: Accidents or disasters from failing structures may result in significant liabilities and hence a systematic analysis should be performed for existing structural items in the system. Routine inspection reports, historical data, documented accidents or problems should be reviewed. Structures such as storage tanks, elevated tanks, reservoirs, bridges, and so on, are major cost items if included as a part of new or upgrade projects. Careful analysis and decisions at the planning stage go a long way with the projects (Almeida & Ramos, 2010).

Aging infrastructure: Apart from reviewing capacities, performance and structural integrity, it is also important to evaluate the age of various infrastructures within the system. Drinking water and wastewater systems or their parts might have been built a long time ago and might require replacement with newer systems. For example: older sanitary systems may develop cracks and cause significant inflow and infiltration issues within the sewer system. This may also result in higher hydraulic loadings at the treatment plants. Since the majority of sewers are below ground, it is difficult to locate and identify each and every malfunctioning item. Hence, sometimes it is economical to replace the old system as a whole considering public health concerns (Bertrand-Krajewski *et al.* 2006; Clark *et al.* 1999; Karpf *et al.* 2011; Kleidorfer *et al.* 2013; Ahmed *et al.* 2011).

Compliance review: Consideration of regulatory issues is critical throughout the planning process (Lombardo, 2004). Regulatory compliance is a key parameter in understanding present system operations. Well operated, maintained and in-compliance systems are healthy signs for infrastructures. Any occasional or frequent violations observed in the past should be highlighted and considered for future improvements. A project's compliance with regulatory programs makes it easier to ensure that the project will not have a significant environmental impact on the areas subject to regulation (Lamont & Young, 2007).

Manpower: A competent, educated and knowledgeable staff is required for system operation and management. Along with the quality of staff, number of employees available per shift or for key tasks should be evaluated. For overseeing operations as well as handling critical events such as accidents, chemical spills, emergency conditions, fires, flooding, and so on, qualified staff are needed (Lindgaard-Jørgensen & Bender, 1994; Malle, 1994; Möderl *et al.* 2008). Hence, it is important that the best staff is hired and nurtured by organizations. Available manpower and staff assessment should be performed during the planning process and future needs identified.

6.4.2 Immediate and future needs

Upon completion of the existing systems review, the next step is to identify present and future needs. Non-compliance issues, capacity problems, malfunctioning items, ageing structures and manpower requirements flagged during existing systems evaluation contribute towards the 'Immediate Needs'. Addressing such issues would help to overcome ongoing drinking water or wastewater infrastructure and management problems. Any 'Immediate Needs' items if not addressed properly would become a problem for future operations. 'Future needs' typically result from changes to environmental laws, regulatory needs, new industrial connections, expected development and growth, increased service demands, and so on, in the area and are briefly discussed below:

Community growth: Community growth is one of the key items that demand more and improved infrastructure to keep up with society and its citizen's demands. Any increase in housing units, restaurants, offices, buildings, conversion of rural or agricultural areas to urban land can result in increased demand for drinking water and wastewater utilities. (Lamont & Young, 2007; Leverenz *et al.* 2011; Lombardo, 2004; Wang Wenyi *et al.* 2014).

Industry connections: Future industrial developments may include construction of new processing industries or expansions and increased production within the existing industries. Processing industries typically have higher water demands and also generate wastewater that needs to be treated. To handle these increased demands, infrastructure upgrades may be needed in the future (Andrews *et al.* 2011; Görgün *et al.* 1999; Pareek, 1992).

Newer regulations: With increasing awareness and concerns for water conservation, its quality and improved sanitation in society, governments and policy makers are implementing more stringent laws for managing water and wastewater systems. Any changes to existing regulations or implementation of newer regulations or laws related to drinking water or wastewater could potentially require infrastructure improvements, hence adding to the future needs (Lamont & Young, 2007; Lombardo, 2004; Válek *et al.* 2011).

6.4.3 Budgetary considerations

Considerable portions of sewer and drinking water infrastructures are owned and operated by government entities and tax-payer's money gets involved with new or upgrade projects (Adams, 2008; Kelman, 2004). Whereas, certain infrastructure utilities are owned by private companies, utility owners, developers, or industries and related projects are sponsored by these private entities. Any changes to 'sewer rates', 'water rates' or related fees can have direct impact on the communities.

With technical, social, political and financial aspects involved, budgetary consideration becomes a key factor that governs implementation of future projects. Some of the important financial considerations during the project planning process may include funds allocation, government incentives for newer and greener technologies, grants and loans, private projects and payback periods.

6.4.4 Priority allocation and recommendations

Upon reviewing existing conditions, immediate needs and future requirements, the next task is developing and proposing solutions. It is important to prioritize recommendations with respect to the community goals, urgent needs, budget availability and regulatory requirements.

The planning process involves screening technologies and management options, evaluating and ranking technical solutions, developing the preferred solution plan, defining an implementation plan, establishing institutional management (such as administration, maintenance, repair or replacement, regulations, etc.), financing, and scheduling (Lombardo, 2004).

6.5 DESIGNING PHASE

The design phase follows the planning process. Commonly, clients select a design services firm, finalize professional fees, terms and conditions, and sign the project contract. The services company can be pursued on a competitive basis such as by soliciting an RFP, RFQ, RFI, and so on (Lamont & Young, 2007; CIPS & NIGP; WSDOT, 2007), based on recommendations by other organizations or contacts, or as a continued service provider firm that has assisted the client with previous projects or the planning process.

Recommendations from the planning study are reviewed, verified or modified if needed and finalized prior to the beginning of design process. During this phase, the services firm typically assists with preparing a facility plan, final design drawings and specifications, obtaining permits, property and easements, and obtaining approvals from regulatory and funding agencies (Lamont & Young, 2007).

Designing is one of the skillful phases requiring involvement of professional experts from several disciplines such as environmental, civil, mechanical, plumbing, electrical, and so on, and a lot of coordination amongst the departments. The designing process can be commonly divided into the preliminary design, final design and construction documentation phase. These phases can vary greatly from project to project, hence general guidelines are provided in this section for understanding the design process.

6.5.1 Project scope and schedule

Finalized scope is also accompanied by project schedule with approximate time frames that include specific dates, deadlines, milestones related to design submissions, permit approvals, land acquisitions, submitting funding requests, bidding and award, construction, start-up, operations, contract close-out, and so on (Lamont & Young, 2007). Schedules are critical and any delays can potentially result in increased costs, liabilities or coordination issues.

6.5.2 Design basis

A well-laid design basis serves as a strong technical background and avoids major hassles down the road during the implementation process. Also, appropriate technologies can be determined based on the design criteria (Lombardo, 2004). The present section discusses the basic factors involved with developing the design basis:

Flows or demands: Flows and demands are crucial for the majority of infrastructure undertakings. There are different ways in which the flows or demands are analyzed while designing utilities or equipment and these may include average, peak, instantaneous, minimum, daily, monthly or annual conditions. The hydraulics associated with the flows is also vital for selecting processes, pipes and equipment (Alaska DEC, 2013; Ohio EPA, 2013; US EPA, 2000d, 2002a–c). For water supply systems, peak consumer demands, fire suppression flows, operating pressures and storage capacities are equally important (Alaska DEC, 2013; USFA, 2008). Normally, drinking water or wastewater requirements are based on the number and types of customers served.

Design loadings: Solids, organic, hydraulic and chemical loadings and associated influent concentrations are vital for designing conveyance and treatment systems. Residential, commercial, industrial, or other types of generation sources can have significant impacts on wastewater characteristics (Lombardo, 2004; Ohio EPA, 2013). Industrial processing facilities connected to sewer systems may discharge specialized chemicals in wastewater and appropriate measures should be taken to make sure that these pollutants can be treated without causing interference with sewage treatment plant operations (OR DEQ). Regarding water treatment, influent turbidity loading, microbial contamination and presence of taste- or odor-producing substances can have potential impacts on treatment process and associated costs (Alaska DEC, 2013; USFA, 2008; WHO and IRC, 2003).

Degree of treatment: Along with flows and loadings, it is important that selected equipment and processes achieve the desired degree of treatment. For water treatment plants, the source type, flow fluctuations and potable water quality requirements decide the complexity of the treatment process. Whereas, the wastewater influent characteristics and effluent permit requirements govern primary, secondary and tertiary processes involved during the treatment process. Proper designs with the expected performance will ensure regulatory compliance (Libabher & Orozco-Jaramillo, 2012; SDWF, Water factsheet; U.S. EPA, 1999a, 2004; U.S. Fire Administration, 2008; van Haandel & van der Lubbe, 2012).

6.5.3 Preliminary design

With the project plan prepared and the basis finalized, the design process begins. Preliminary designs in many cases have the flexibility of comparing and evaluating various design components, reviewing available options and making appropriate selections suitable for the project. Professional experience, skills and knowledge come in handy for analyzing and choosing proper economical alternatives for satisfying project needs. Various aspects involved with the preliminary design phase are discussed below:

Land acquisition: The land acquisition process involves obtaining control of the sites needed for the project through purchase, lease, or easements. Availability of potential project sites is critical to the success of the project. The land acquisition process, especially for critical project components, should begin at a very early stage in the project. The client, with assistance from an attorney, negotiates with landowner to secure options or a purchase agreement on the land. (Lamont & Young, 2007). Evaluating the site in terms of its environmental conditions, physical features, and characteristics provides the information needed to size, select, and locate the appropriate drinking water or wastewater system (U.S. EPA, 2005a,b). Land availability and footprint requirements decide the future course of the project design.

Technology selection: Drinking water and wastewater industries are evolving rapidly with the emergence of newer processing technologies on a regular basis. For example, several wastewater treatment technologies such as the activated sludge process, membrane bioreactor, sequencing batch reactor, oxidation ditch, lagoon, and so on, are available in the market with each having their own advantages and disadvantages (U.S. EPA, 2006b, 2013). Hence, comparison and selection of appropriate technology is an important aspect of the design process (Lombardo, 2004).

Capital costs: Capital costs are crucial for project economics and many times are site specific, dependent upon local requirements, selected application and economy of scale (U.S. EPA, 2013). Hence, while choosing any unit processes, operations, technologies or even the design approach, financial aspects should be evaluated as well. It is possible that certain alternatives are technically sound, but might incur higher costs. The best suitable and cost effective alternatives can be identified by contacting relevant manufacturers, vendors or suppliers and comparing available technologies.

Operations costs: Technological solutions may have different capital costs and operations and maintenance (O&M) costs. Hence for a fair comparison, present worth life-cycle cost (LCC) analysis should be performed for different alternatives prior to selection. LCCs are the capital cost estimate for the solution plus the future annual O&M costs discounted to their present value, less the present value of salvage at the end of the analysis period. This evaluation will be helpful in offering economic choices to the clients while maintaining project integrity (Jones *et al.* 2013; Lamont & Young, 2007; Lombardo, 2004).

Project documents: Preliminary design documents typically include final design criteria, preliminary drawings and outline specifications (Lamont & Young, 2007). Once alternatives are selected for the project, equipment drawings, cut-sheets, manuals, design calculations are requested from the respective manufacturers and sales representatives. If the equipment are supplied by different sources, designers should ensure that they are compatible with each other. Preliminary drawings such as process flow schematics, hydraulic profiles, existing and proposed site layout, demolition drawings, equipment details, and so on, are prepared during this phase. Many times, the preliminary drawings focus on process aspects and other disciplines such as electrical, mechanical, plumbing, and so on, have a limited role at this stage. Project preliminary specifications supporting the drawings are also developed for the client's review.

Project estimates: Along with the preparation of preliminary drawings and specifications, preliminary estimates of total project costs are prepared. In addition to the capital and O&M costs for selected technologies, the total project costs may also include costs for permits, bonds and insurances, traffic management, mobilization and demobilization, process buildings, excavation, installation, commissioning, and start-up, and so on. Being at the preliminary design stage, these estimates typically involve contingencies to take into account any unknown or unforeseen conditions. Comparing with planning phase cost estimates, project price changes or scope revisions are also made. Cost estimates are submitted to client for review (Commonwealth of Massachusetts, 2006; Lamont & Young, 2007; Lombardo, 2004; WSDOT, 2009).

Permit requirements: Regulatory and permit requirements are vital for project execution. Permits are usually applied for early in the design phase since many of the permit applications require at least preliminary design drawings, or at a minimum, that the actual location of project components is known. Many agencies take a preliminary look at the submittals and request additional information before making comments. Regulatory agency comments are incorporated in design drawings and resubmitted. During design, list of permits and approvals should be updated and used to track the status of permit applications. (Lamont & Young, 2007). All the necessary permits should be thoroughly reviewed during preliminary design.

Client review: Preliminary design documents including design basis, project drawings, specifications, permit applications, grant or loan options and opinion on project construction costs are typically submitted to the client. Feedback or recommendations received are used for final design preparations.

6.5.4 Final design

Final design documents are detailed and involve inputs from various disciplines. Following are general guidelines towards completion of the documents, however, it should be noted that this design phase can vary substantially from project to project.

Drawings: The finalization process involves significant technical inputs and inter-departmental coordination. Detailed drawings may include professional seals, to-the-scale drawings, location maps, plans and profiles, equipment details, construction notes, permit specific information, civil and structural drawings, electrical and instrumentation drawings, mechanical and plumbing drawings, and so on (Commonwealth of Massachusetts, 2006; Lamont & Young, 2007; Li, 1992).

Specifications: Drawings are supported by project specifications. These documents provide specific information about selected equipment and processes such as manufacturer's details, materials of construction, standards, warranties, past installations, performance design criteria, construction methods, testing, start-up procedures, accessories, instrumentation, electrical requirements, training information, inspection, and so on. These specifications along with drawings become the basis for the project bidding documents for construction purposes (Brepols, 2011; Commonwealth of Massachusetts, 2006; Lamont & Young, 2007).

Final permits: Comments received from permitting agencies during the permit review and approval process get incorporated in the final design. Final plans and specifications, when complete, are then submitted to the regulatory and funding agencies to have project plan approval. Plan approval is required prior to award of contracts (Lamont & Young, 2007). Even though the design is foolproof, if proper permits are not obtained in time, the project may face major hurdles or delays. Professionals should keep themselves on top of the permitting process by submitting updated forms in time, providing application fees and keeping a track of correspondence with the approving authority.

Construction sequence: The drinking water and wastewater industry is a service sector and community members heavily rely on water supply as well as sanitation systems. Any service interruptions can cause major inconvenience to society. Hence, proposed projects should proceed smoothly without affecting the ongoing activities. Considering the complex nature of projects, a well-developed construction sequence is necessary to ensure uninterrupted operations, maintain water quality, regulatory compliance and hygienic conditions (Commonwealth of Massachusetts, 2006; WSDOT, 2007, 2009).

Final project costs: Once final drawings and specifications are prepared and approved by appropriate permitting authorities, a clear idea about the project is obtained. This is a good time to revise and update the preliminary project cost estimate developed earlier. Several unknowns can be eliminated at this point and more precise item costs can be added. Lesser contingencies are needed to the estimates and more reasonable project costs can be prepared. Appendices N and O include sample cost estimation worksheets for drinking water and wastewater projects (Commonwealth of Massachusetts, 2006; Lamont & Young, 2007; WSDOT, 2007, 2009).

Client approval: Just like the preliminary phase, final drawings, specifications, approved permits, cost estimates and other related documents are submitted to the client for review. Once the client gives the 'go-ahead', preparations for the construction process begins.

Safety factor: Realizing the unknowns and any unforeseen situations during construction work, the designs may involve certain 'factors of safety'. This process helps to minimize major 'change orders' during construction work and any potential conflicts between the parties involved, providing smoother operations during the project's execution. However, excessive safety factors can sometimes deviate from project goals resulting in higher costs and unexpected system performances. This is an area where past experience and education comes into the picture. Professionals should make reasonable choices when a factor of safety is needed.

6.6 BIDDING AND CONSTRUCTION PHASE

Projects can proceed in many ways from the final design stage to the construction phase and the discussions below are intended to provide a general understanding about this topic and should be used for guidance purposes only:

Bid advertisements: Once the bid documents are ready, projects are typically advertised in public. When tax-payers' money is involved, solicitations provide a good opportunity for the community as well to understand proposed drinking water or wastewater projects by governments in their region. Advertisements may be published in newspapers, government websites or with other relevant media outlets (Lamont & Young, 2007; CIPS & NIGP; WSDOT, 2007, 2009).

Projects bidding: During the competitive bidding process, bid packages and costs submitted by interested parties are opened publicly and construction firm or firms get selected for project execution, often the lowest bidder. Sometimes the work is awarded to a single contractor, however, in certain cases the client appoints a general contractor responsible for execution of overall work and there can be other sub-contractors handling specific portions of the project. Contracts are signed with the selected firm or firms for moving forward with the implementation process (Commonwealth of Massachusetts, 2006; Lamont & Young, 2007; CIPS & NIGP; WSDOT, 2007, 2009).

Contractor selection: Each project is unique and its execution depends upon client type, funding agencies involved, project size, expected timeline, social constraints, and other key factors. Based on project needs, the contracting firm may get selected in different ways. As discussed earlier, the client can go for a competitive bidding process, may choose from a pool of pre-qualified companies, or can utilize a firm that has worked on previous projects to hand over the construction work at an agreeable cost.

Submittals review: Administrative, product and equipment information must be submitted by the contractor to the engineer for review and approval. Administrative submittals typically include a list of suppliers, subcontractors and others that the general contractor intends to use on the project, the proposed construction schedule, and affirmative action requirements. Product and equipment submittals allow the engineer to compare the products and equipment the contractor proposes to use to the items called for in the specifications. The submittals give the engineer a chance to review and approve the proposed products and equipment before they are purchased and incorporated into the work (Lamont & Young, 2007). Upon approval, the contracting firm starts ordering equipment and materials to be shipped to the site.

Construction coordination: A lot of coordination is involved during the project execution stage. Depending upon the scope, several entities might get involved, such as contractor, sub-contractors, client, design firm, permitting authority, local governments, transportation department, owners of other utilities located within the project area and anyone else affected by the work. Coordination may also include managing workers, ensuring workplace safety, erosion and sediment control provisions, supervising mobilization and demobilization of equipment, handling emergency situations, traffic controls, and so on. The construction management team needs to ensure that the proposed 'Sequence of Construction' is strictly followed so that operation of other facilities and systems remain unaffected. This avoids any sudden interruption in drinking water or sewer services to the community and minimizes local issues. Appendices B and L include a glossary of construction management related terminologies and a listing of common heavy equipment for reference purposes, respectively.

Project supervision: With major liabilities and responsibilities involved during construction work, specialized professionals are required to oversee project activities. Project supervision involves tasks such as project schedule management, attending meetings with authorities, contractors and engineers, conducting onsite observations, keeping a daily observation log, verifying testing and start-ups, updating record drawings, reviewing payment requests, preparing a list of items requiring completion or correction (punch list) and final review of work, and so on (Lamont & Young, 2007).

Substantial completion: When the project is ready for its intended use, the contractor notifies relevant parties (such as client, government and engineer). After inspection by the approving agencies and satisfaction of project status, the Certificate of Substantial Completion is issued (Lamont & Young, 2007).

Testing and start-up: Upon substantial completion, the drinking water or wastewater systems are tested and put into operation during the process start-up phase. Components such as intake structures, pumps, flow meters, water or wastewater processing equipment and tanks, the disinfection system, solids handling, the aeration system, the chemical feed system, electrical controls, the emergency generator, and other system components are tested. Testing for the collection system or distribution piping typically include low-pressure air testing, manhole vacuum testing, pump station run testing, fire hydrant testing, and pipeline deflection testing. Usually the systems are tested with clean water initially. After testing and start-up, the system is drained of clean water and raw sewage (in the case of the wastewater system) or raw intake water (for the drinking water system) is put into service. The wastewater treatment plant may require a seeding process (addition of septage waste or sludge from another facility) to begin and maintain the biological treatment process until process control can be optimized (Alaska DEC, 2013; Lamont & Young, 2007).

Contract closeout: As the project reaches the completion stage, final close-out documents are typically submitted to the regulatory and funding agencies which may include operation and maintenance manuals, recorded drawings, claims statements, an affidavit of payment of debts and claims, a contractor's release of liens and consent of surety to final payment, warranties, certificates of inspections, testing and start-up field reports, a final inspection report with final punch list items, if necessary and a final payment request. Once the regulatory and funding agencies are satisfied that the project is complete, the funding agencies release the contractors' remaining retainage (Lamont & Young, 2007).

6.7 SYSTEM OPERATIONS AND MANAGEMENT

Operational and management procedures can vary significantly from system to system and location to location. The present section provides an overview of typical duties performed by operations and support staff (Alaska DEC, 2013; Ohio EPA, 2013; Oregon DEQ, 2013; U.S. EPA, 2000a–e, 2005, 2006; WS DOE, 2010).

Routine site visits: Operational staff schedule routine site visits to various sections and portions of the systems to make sure that all the processes are in order. Alarms, sirens, flashing lights, emergency signals, leaks, noises, vibrations, excessive heat generation, and so on, are important visuals during the inspection visits. Any abnormal observations should be noted down and rectified as soon as possible. System violations should be reported promptly. Perform routine security inspections.

Operations: Duties may include overseeing system operation efficiencies and effectiveness of treatment processes, recording meters and gauge readings, log inflows and water production rates, water sampling, preparing operational parameters log-sheets, performing field duties, loading and unloading chemicals, installation of water connections, hydrants and valves, establishing traffic control devices, sludge removal and hauling, and so on.

Maintenance: Tasks include cleaning, flushing, oiling, greasing, lubrication, repairing, painting, calibration, start-up, shut-down, backwashing, blow-down, tests, troubleshooting, corrosion remediation, and so on. In addition, operators need to keep updated the water or sewer system drawings, plans, maps, specifications, maintenance tools, system security services, and so on.

Lab operations: Laboratory analysis is an integral part of both water and wastewater systems. The tasks may include sampling, shipping, storage and analysis of various process parameters. Some of the important wastewater system related analytical parameters include BOD, COD, suspended solids, ammonia, nitrate, nitrite, phosphorus, pH, temperature, total coliform bacteria, alkalinity, dissolved oxygen, residual chlorine, and so on. Similarly, common drinking water system related parameters include color, odor, temperature, pH, alkalinity, conductivity, taste, chloride, chromium, fluoride, hardness, turbidity, dissolved solids, residual chlorine, and so on. In addition, lab operators may perform bench-scale or lab-scale studies for process improvements or system review. Laboratory operations include lab equipment, instruments, glassware, safety gears, safety equipment, chemical supplies, reagents, and so on. Analysis can be performed at the treatment facility site or samples can be shipped to third party laboratories specializing in water and wastewater systems. Appendices C, G, K and P include reference databases for laboratory related terminologies, formulae, an inventory list and cost estimation respectively.

Permits and reports: Maintaining permit records and compliance activities, violations logs, preparing water quality reports, analytical results, monthly and annual operations reports, sludge management datasheets, annual budgets, equipment and supplies records, system maintenance plans, sampling plan, standard operating procedures (SOPs), consumer confidence reports, consumer complaint records, documentation of current environmental laws and regulations, and so on, are important tasks for the operators and supporting staff.

Inventory management: Performing inventory management of process chemicals, instruments, equipment, control systems, recorders, meters, pipes, valves, fittings, maintenance tools, laboratory supplies, other spare parts.

Safety measures: Inspecting and maintaining safety equipment (such as fire extinguishers, first-aid kits, eye washes, showers, fire hydrants, etc.), electricity backup systems (such as emergency generators), facility illumination devices, security mechanisms; developing an emergency response plan, chemical spills action plan; maintaining an emergency contact list, preparing for any unforeseen situations and accidents; educating others with emergency procedures, and so on.

Other duties: Staff management, employee records, timesheets, utility bills, paychecks, purchase orders, vouchers, budget preparation, professional certifications, administrative supplies, handling and responding to public queries, sending fliers and important notices to customers, attending training sessions and conferences, participating in owner and board meetings and public relations programs, obtaining approvals, conducting plant tours, and so on, are some of the other duties involved.

Chapter 7

Realizing exceptional achievements

A professional's journey in the drinking water and wastewater field begins as a novice student followed by entry-level professional and all the way to the well-established expert position. Irrespective of the job profile, individuals strive for excellence, respect and recognition in the field. What really makes some people exceptional achievers amongst others is their ability for taking the right decisions at the right time, thinking out of the box, taking reasonable risks and going one step beyond what others have achieved in the field with devotion, dedication and determination. An exceptional performer makes the best of the available resources and situations. Achievers become trend setters and an inspiration for others in the field. Their work is remembered by the next generations and experiences are referred to for motivation. Present section deals with the exceptional abilities of professionals and the path to achieve this greatness. Since the focus of the book is on drinking water and wastewater professionals, this topic is discussed keeping in mind industry specific achievements.

7.1 REALIZING EXCEPTIONAL GOALS

Once the environmental professionals embark on their journey in the drinking water and wastewater field, they should start aiming for exceptional goals that are beyond the day-to-day routine of work-life. As discussed in Chapter 4, these exceptional achievements can be envisioned in several ways such as the leader of national organization, successful entrepreneur, renowned author of important publications such as technical books, standards or manuals, an extraordinary researcher, developer of new patents, contributor to visionary documents, innovator of ground-breaking technologies, key member on technical committees, and many more. These goals will help one expand his or her professional horizons, rise beyond everyday practice and become a trend setter in the industry. Such achievements are naturally followed by name and fame with recognition amongst peers. The earlier the better; professionals can go a long way if such aims are realized at the beginning of their careers.

7.2 MAKING PROFESSION A PASSION

To be an exceptional practitioner, individual should be highly motivated for the work one is doing and make it a passion. There are several ways to develop this appetite for the drinking water or wastewater profession. A professional needs to nurture the habit of asking oneself questions such as what, why, when, how while choosing the profession, performing tasks and by justifying steps and decisions taken at each stage with the appropriate thought process. This way, professionals will start inculcating confidence and develop a clear idea about the profession and personal ambitions. Thoughts followed by actions will help to substantiate this professional approach for extraordinary achievements.

7.3 CREATIVITY AND REASONABLE RISKS

There are many projects, tasks or decisions that drinking water and wastewater professionals come across for the first time in their professional journey. Restricting oneself from taking conscious decisions and avoiding calculated risks can sometimes keep the individual away from creativity. Hence, it is important to take logical decisions at various stages of life that one thinks are appropriate and beneficial for personal growth as well as society.

7.4 ACHIEVEMENTS WITH ETHICS

Definition of professional achievements in life can vary from person to person. Some consider money as the parameter to measure success, whereas others consider executive positions, experience, publications, groundbreaking work, and so on. Although these are important aspects, irrespective of such priorities, professionals should make every attempt to achieve these goals with ethics and principles. To be recognized amongst peers, respected by experts and gaining credibility in the drinking water or wastewater field, work ethics play a vital role and additionally can help to further boost and glorify the monetary or other professional achievements.

7.5 STRUGGLE AGAINST THE ODDS

No world or profession is perfect and just being a good academician, researcher, designer, engineer, consultant or operator is not enough to develop oneself as an extraordinary professional. Despite every attempt at sincere work, dedication and commitment, there are always unfamiliar situations for the professionals at work place and within the society. It is important to realize that to become an extraordinary professional, one needs to rise beyond all the odds, expand vision, think globally and motivate others, as well, to achieve greater professional heights in the field.

7.6 APPRECIATING THE NOBLE PROFESSION

Water is one of the basic needs for the existence of life. It is difficult to imagine the environment without water. For the survival and advancement of civilizations throughout the world, it is important that available water sources can sustain growing demands, societies can be served with high water quality and improved sanitation is provided to maintain hygienic conditions and better public health. For these reasons, the drinking water and wastewater field stands out amongst other professions. Even today, in many parts of the globe, the environmental field is still not a priority. Hence, professionals have a moral responsibility to make conscious efforts for global improvements and spread the importance of this field by taking smart decisions and providing reasonable solutions. The journey in the drinking water and wastewater field can not only serve as a noble profession but also provide a satisfying experience with direct contributions towards the betterment of society and the world as a whole.

7.7 AN INSPIRATION FOR THE NEXT GENERATION

The upcoming generations are smart, quick learners, full of enthusiasm and ready to establish themselves in the society. As members of the drinking water and wastewater community, they look forward to guidance from peers and can themselves become the pioneers of groundbreaking work if given the right directions. As an extraordinary professional, one needs to inculcate habits of motivating budding professionals and letting them understand and realize the importance of this field. No achievement can be higher than becoming an inspiration for the next generation of drinking water and wastewater professionals and for the sake of civilization.

References

Adams W. B. (2008). Major Water Industry Reform in Australia: A Case Study from Queensland. IWA Publishing, UK, Water Practice & Technology.

Ahmed Z., Main D., Davies Murphy K. and Petelka L. (2011). Process benchmarking a Balanced-Scorecard system: developing best practices for Abu Dhabi sewer expansion. *Water Utility Management International*, **6.2**, 25.

Alaska Department of Environmental Conservation (Alaska DEC), Division of Water (2013). Introduction to Small Water Systems Course Manual. http://dec.alaska.gov/water/opcert/IntroSmallCorrOnlineCourses.html. (accessed 28 July 2014).

Alegre H., Hirner W., Baptista J. M., Parena R., Cubillo F., Cabrera E. Jr., Merkel W., Duarte P. (2006). Performance Indicators for Water Supply Services. 2nd edn, IWA Publishing, UK.

Almeida A. B. and Ramos H. M. (2010). Water supply operation: diagnosis and reliability analysis in a Lisbon pumping system. *Journal of Water Supply: Research and Technology – AQUA*, **59**(1), 66–78.

Ambulkar A., Zeller S. N. and Klinger D. (2011). One-year operational data for Wiconisco Wastewater Treatment Plant illustrating process improvements and carbon footprint reduction with SolarBee® circulators. *Water Practice & Technology*, **6**(1).

American Geophysical Union (AGU) (2001). Put Your Science to Work, the Take-Charge Career Guide for Scientists by P. S. Fiske, www.agu.org/books/sp/v053/SP053.pdf (accessed 4 August 2014).

Andrews M., Berardo P. and Foster D. (2011). The sustainable industrial water cycle – a review of the economics and approach. *Water Science & Technology: Water Supply*, **117**(1), 67–77.

Arnbjerg-Nielsen K., Willems P., Olsson J., Beecham S., Pathirana A., Bülow Gregersen I., Madsen H. and Nguyen V.-T.-V. (2013). Impacts of climate change on rainfall extremes and urban drainage systems: a review. *Water Science & Technology*, **68**(1), 16–28.

Arthur S., Ashley R. M. and Nalluri C. (1996). Near bed solids transport in sewers. *Water Science and Technology* **33**(9), 69–76.

Association of American Colleges and Universities (AACU) (2007). College Learning for the New Global Century, A Report from the National Leadership Council for Liberal Education and America's Promise. http://www.aacu.org/leap/documents/GlobalCentury_final.pdf (accessed 4 August 2014).

Banasiak R. (2008). Hydraulic performance of sewer pipes with deposited sediments. *Water Science & Technology – WST*, **57**(11), 1743–1748.

Barry J. P., Q.C. and LeBlanc J. G. (2007). Legal consequences of data reporting and public perception: pitfalls of the regulatory system in municipal wastewater treatment facilities' pursuit of environmental advancement. *Water Practice & Technology*, **2**(4), 1–8.

Bassan M., Mbéguéré M., Tchonda T., Zabsonre F. and Strande L. (2013). Integrated faecal sludge management scheme for the cities of Burkina Faso. *Journal of Water, Sanitation and Hygiene for Development,* **3**(2), 216–221.

Bell K. Y. and Abel S. (2011). Optimization of WWTP aeration process upgrades for energy efficiency. *Water Practice & Technology,* **6**(2). http://www.apg-neuros.com/Uploads/Documents/7%20-%20Aeration%20 Optimization%20-%20IWA%20Publishings%20-%202011.pdf (accessed 28 July 2014).

Benedetti L., Langeveld J., Comeau A., Corominas L., Daigger, G., Martin C., Mikkelsen P. S., Vezzaro L., Weijers S. and Vanrolleghem P. A. (2013). Modelling and monitoring of integrated urban wastewater systems: review on status and perspectives. *Water Science & Technology,* **68**(6), 1203–1215.

Bertrand-Krajewski J.-L., Cardoso M. A., Ellis B., Frehmann T., Giulianelli M., Gujer W., Krebs P., Pliska Z., Pollert J. and Pryl K. (2006). Towards a better knowledge and management of infiltration and exfiltration in sewer systems: the APUSS project. *Water Practice & Technology,* **1**(1).

Brandt M., Middleton R., Wheale G. and Schulting F. (2011). Energy efficiency in the water industry, a Global Research Project. *Water Practice & Technology,* **6**(2).

Breach B. (2012). Drinking Water Quality Management from Catchment to Consumer: A Practical Guide for Utilities Based on Water Safety Plans. IWA Publishing, UK.

Brepols C. (2011). Operating Large Scale Membrane Bioreactors for Municipal Wastewater Treatment. IWA Publishing, UK.

Brocard D., Garcia M., Kunetz T., Sobanski J., Waratuke A. and Weber T. (2009). Physical Model Study of Chicago's New Calumet Influent Pumping Station. *Water Practice & Technology,* **4**(1), 1–9.

Brown M., Barley B. and Wood H. (2002). Minewater Treatment – Technology, Application and Policy. IWA Publishing, UK.

Burke V., Duennbier U. and Massmann G. (2013). The effect of aeration on the removal of wastewater-derived pharmaceutical residues from groundwater – a laboratory study. *Water Science & Technology,* **67**(3), 658–666.

California State University, Sacramento, California, Office of Water Programs (CSU OWP). USA Glossary section. http://www.owp.csus.edu/glossary/index.php (accessed 24 July 2014).

Chartered Institute of Purchasing & Supply (CIPS) and The Institute for Public Procurement (NIGP). Partnership for Public Procurement. Glossary – Public Procurement Dictionary of Terms. www.globalpublicprocurement.org/ Resources/Glossary (accessed May 2014).

Clark R. M., Rizzo G. S., Belknap J. A. and Cochrane C. (1999). Water quality and the replacement and repair of drinking water infrastructure: the Washington, DC case study, IWA Publishing, UK. *Journal of Water Supply: Research and Technology – AQUA*, **48**, 106–114.

Computing Research Association's Committee on the Status of Women in Computing Research (CRA-W) (2000). Graduate School Information Guide Guidance for Undergraduate and Graduate Students. http://www.cra.org/ uploads/documents/resources/student/grad-guide.pdf (accessed 4 August 2014).

Commonwealth of Massachusetts, Division of Capital Asset Management (February 2006). Consultants Estimating Manual. http://www.mass.gov/anf/docs/dcam/dlforms/cem-feb06.pdf (accessed 30 July 2014).

Duke University, Summary of Academic Titles, Provost's Office of Faculty Affairs. www.facultyaffairs.provost. duke.edu/titles.html (accessed 4 August 2014).

European Commission (EU) (1999) Proceedings of the Workshop on 'Problems around sludge', 18–19 November 1999, Stresa (Italy), Session 3: Technology and innovative options related to sludge management. http://ec.europa. eu/environment/waste/sludge/pdf/workshoppart4.pdf (accessed 6 August 2014).

Farley M. and Trow S. (2003). Losses in Water Distribution Networks: A Practitioner's Guide to Assessment, Monitoring and Control. IWA Publishing, UK.

Fattal B., Eisawy A. M., Dotan A., Shuval H. I. and Mancy K. H. (1989). Impact of water quality on fish production based on Egyptian and Israeli practices. *Water Science & Technology,* **21**(3), 27–33.

Gamberi C. and Scholes D., National Postdoctoral Association (NPA). Going In With Your Eyes Open: What To Ask Before Accepting A U.S. Postdoctoral Position. http://www.nationalpostdoc.org/graduate-students-29 (accessed 4 August 2014).

Gangl G., Ertl T., Kretschmer F., Kainz H. and Haberl R. (2007). Development of a guideline for sewer operation and maintenance in Austria. *Water Practice & Technology*, **21**(1).

Geller Y., Zimmerman R. and Friedler E. (2007). Evaluation of the hydraulic performance of the regional sewer network of the greater Tel-Aviv Metropolitan Area (Israel). *Water Practice & Technology,* **2**(2).

Gökkus Ü. (1995). Computer-aided design of submarine pipelines. *Water Science and Technology*, **32**(2), 233–240.

Görgün E., Germirli Babuna F., Orhon D., Özbaşaran M. and Seçkin N. (1999). Wastewater characterization in metropolitan areas with significant agro industries. *Water Science and Technology,* **40**(1), 13–21.

Hahn H. H., Hoffmann E., Kleinschmidt, A. and Klute, R. (1990). Upgrading a large municipal sewage treatment plant by a combination of biological and chemical processes. *Water Science & Technology*, **22**(7–8), 105–112.

Hahn H. H., Hoffmann E. and Odegaard H. (2004). Chemical Water and Wastewater Treatment VIII. IWA Publishing, UK.

Harremös P., Andersen H. S., Dupont R., Jacobsen P. and Rindel K. (2002). Analysis of scenarios for sewerage, wastewater treatment and prioritised load on environment from the Greater City of Copenhagen. *Water Science & Technology,* **45**(3), 95–100.

Heard T. R., Hoyle B. G. and Hieatt M. J. (2002). Aerated biological filtration for the removal of ammonia and manganese in a major new water treatment works under construction in Hong Kong. *Water Supply,* **2**(1), 47–56.

Jones C. J., Laky D., Galambos I., Avendano C. and Colvin V. L. (2013). Life cycle analysis of two Hungarian drinking water arsenic removal technologies. *Water Science & Technology: Water Supply.*

Karpf C., Hoeft S., Scheffer C., Fuchs L. and Krebs P. (2011). Groundwater infiltration, surface water inflow and sewerage exfiltration considering hydrodynamic conditions in sewer systems. *Water Science & Technology,* **63**(9), 1841–1848.

Kazner C., Wintgens T. and Dillon P. (2012). Water Reclamation Technologies for Safe Managed Aquifer Recharge. IWA Publishing, UK.

Kelman J. (2004). Effective subsidies in developing countries. *Water Science & Technology,* **49**(7), 55–59.

Kleidorfer M., Möderl M., Tscheikner-Gratl F., Hammerer M., Kinzel H. and Rauch W. (2013). Integrated planning of rehabilitation strategies for sewers. *Water Science & Technology,* **68**(1), 176–183.

Lamont H. and Young K. (2007). Guide to Developing a Municipal Wastewater Project – for small, rural communities in New York State. Lamont Engineers, New York, USA. http://www.lamontengineers.com/documents/GuidanceManual.pdf (accessed 30 July 2014).

Laureano P. (2010). Ancient water management techniques to counteract drought and desertification in the Mediterranean. *Water Science & Technology: Water Supply*, **10**(4), 495–503.

Leverenz H. L., Tchobanoglous G. and Asano, T. (2011). Direct potable reuse: a future imperative. *Journal of Water Reuse and Desalination*, **1**(1), 2–10.

Li X.-W. (1992). Computer-aided design for stabilization ponds. *Water Science & Technology*, **26**(9–11), 2425–2428.

Libabher, M. and Orozco-Jaramillo A. (2012). Sustainable Treatment and Reuse of Municipal Wastewater. IWA Publishing, UK.

Lindgaard-Jørgensen P. and Bender, K. (1994). Review of environmental accidents and incidents. *Water Science & Technology*, **29**(3), 165–172.

Lombardo P. (2004). Cluster Wastewater Systems Planning Handbook. *Project No. WU-HT-01-45*, Prepared for the National Decentralized Water Resources Capacity Development Project, Washington University, St. Louis, MO, by Lombardo Associates, Inc., Newton, MA. http://www.ndwrcdp.org/documents/WU-HT-01-45/WUHT0145_web1.pdf (accessed 6 August 2014).

Malle K.-G. (1994). Accidental spills – frequency, importance, control, countermeasures. *Water Science & Technology*, **29**(3), 149–163.

Manahan C. L., National Postdoctoral Association (NPA). Finding the 'Perfect Postdoc' for You. http://www.nationalpostdoc.org/graduate-students-29 (accessed 4 August 2014).

Marchiso S. (2000). Sustainable management of water resources and international law. *Water Science & Technology*, **42**(1–2), 241–247.

Merrett S. (2007). The Price of Water – Study in Water Resource Economics and Management. 2nd edn. IWA Publishing, UK.

Milan D., Heinz-Jurgen Brauch and Michael C. Kavanaugh (2008). Groundwater Management in Large River Basins. IWA Publishing, UK.

Möderl M., Vanham D., De Toffol S. and Rauch W. (2008). Potential impact of natural hazards on water supply systems in Alpine regions. *Water Practice & Technology*, **3**(3), 1–8.

National Endowment for Financial Education (NEFE) (2011). Right on the Money Financial Know-How for Financial Freedom, 2nd edn. http://www.dfi.wa.gov/consumers/education/college_book/collegebook.pdf (accessed 4 August 2014).

National Postdoctoral Association (NPA). Definition of Postdoc Scholars. http://www.nationalpostdoc.org/policy-22/what-is-a-postdoc (accessed 4 August 2014).

National Science Board (NSB) (2014). Science and Engineering Indicators 2014. Arlington, VA: National Science Foundation (NSB 14-01). http://www.nsf.gov/statistics/seind14/ (accessed 4 August 2014).

Nebraska Department of Economic Development (NEDED). Employer's Guidebook to Developing a Successful Internship Program. The Intern Nebraska Program. http://neded.org/files/businessdevelopment/internne/EmployerGuidebook_DevelopingSuccessfulInternshipProgram.pdf (accessed 10 August 2014).

North Central Regional Educational Laboratory (NCREL) (1999). Professional Development: Learning from the Best – A Toolkit for Schools and Districts Based on the National Awards Program for Model Professional Development by Emily Hassel. http://www.learningpt.org/pdfs/pd/lftb.pdf (accessed 19 August 2014).

Nuorteva P., Keskinen M. and Varis O. (2010). Water, livelihoods and climate change adaptation in the Tonle Sap Lake area, Cambodia: learning from the past to understand the future. *Journal of Water and Climate Change*, **1**(1), 87–101.

Ohio Environmental Protection Agency (Ohio EPA), Division of Surface Water (2013). Design Standards: Wastewater Treatment Plants & Collection Systems ≤ 100,000 gpd Green. www.epa.ohio.gov/Portals/35/documents/greenbook_2013.pdf (accessed 7 August 2014).

Oliveira S. C. and von Sperling M. (2011). Performance evaluation of different wastewater treatment technologies operating in a developing country. *Journal of Water, Sanitation and Hygiene for Development*, 37–56.

Olsson G. (2007). Strategies to close water supply and demand gap. *Water Science & Technology: Water Supply*, **7**(4), 103–110.

Pareek, N. K. (1992). Industrial Wastewater Management in Developing Countries. *Water Science & Technology*, **25**(1), 69–74.

Park Jaeyoung, Oh Jin Hwan, Evans E. A., Lally, M. F., Hobson K. L. and Ellis T. G. (2012). Industrial wastewater treatment by on-site pilot static granular bed reactor (SGBR). *Water Practice & Technology,* **7**(1).

Plastic Pipe Institute Inc. (PPI), USA. Handbook of PE Pipe. http://plasticpipe.org/pdf/glossary.pdf (accessed 24 July 2014).

Pollert J., Ugarelli R., Saegrov S., Schilling W. and Federico V. Di. (2005). The hydraulic capacity of deteriorating sewer systems. *Water Science & Technology*, **52**(12), 207–214.

Rieu-Clarke A. (2005). International Law and Sustainable Development. IWA Publishing, UK.

Rochmadi R., Ciptaraharja I. and Setiadi, T. (2010). Evaluation of the decentralized wastewater treatment plants in four provinces in Indonesia. *Water Practice & Technology*, **5**(4), 1–21.

Rosemarin A., Ekane N., Caldwell I., Kvarnstrom E., McConville J., Ruben C. and Fogde M. (2008). Pathways for Sustainable Sanitation – Achieving the Millennium Development Goals. IWA Publishing, UK, SEI/IWA, 56p.

Rouse M. (2013). Institutional Governance and Regulation of Water Services – The Essential Elements Book, 2nd edn, IWA Publishing, UK.

Rousseau P. and Tranchant E. (2002). Business partners for development-KwaZulu-Natal pilot project, South Africa: a new approach to the delivery of sustainable water and sanitation services in poor peri-urban areas. *Water Supply*, **2**(4), 197–202.

Safe Drinking Water Foundation (SDWF), Canada. Wastewater Treatment Factsheet. www.safewater.org/resources/fact-sheets.html (accessed 30 July 2014).

Safe Drinking Water Foundation (SDWF), Canada. Conventional Water Treatment: Coagulation and Filtration Factsheet. www.safewater.org/resources/fact-sheets.html (accessed 30 July 2014).

Schreiner B. and van Koppen B. (2003). Policy and law for addressing poverty, race and gender in the water sector: the case of South Africa, IWA Publishing, UK. *Water Policy*, **5**(5), 489–501.

Shamos Michael I. (2002). Handbook of Academic Titles. Carnegie Mellon University. http://euro.ecom.cmu.edu/titles/titlebook.htm (accessed 4 August 2014).

Smakhtin V. U. and Schipper E. L. F. (2008). Droughts: the impact of semantics and perceptions, IWA Publishing, UK. *Water Policy*, **10**(2), 131–143.

Sonntag C. von and Gunten U. von. (2012). Chemistry of Ozone in Water and Wastewater Treatment: From Basic Principles to Applications. IWA Publishing, UK.

Sousa J., Ribeiro A. and da Conceição Cunha Maria and Antunes António (2002). An optimization approach to wastewater systems planning at regional level. *Journal of Hydroinformatics*, 115–123.

State of Oregon Department of Environmental Quality (OR DEQ). Industrial Pretreatment Program, Introduction Section. http://www.deq.state.or.us/wq/pretreatment/intro.htm (accessed 14 August 2014).

State of Oregon Department of Environmental Quality (Oregon DEQ), Laboratory and Environmental Assessment Division, USA (2013). Quality Manual. http://www.deq.state.or.us/lab/techrpts/docs/DEQ91LAB0006QMP.pdf (accessed 25 July 2014).

Suzuki Y. and Minami, T. (1991). Technological development of a wastewater reclamation process for recreational reuse: an approach to advanced wastewater treatment featuring reverse osmosis membrane. *Water Science & Technology*, **23**(7–9), 1629–1638.

Tennessee Department of Environment and Conservation (TN DEC), Fleming Training Center, Division of Water Resources (2011). Wastewater Treatment Mathematical Formulas. 2nd edn, http://www.tn.gov/environment/water/fleming_math-books.shtml (accessed 24 July 2014).

The Sloan Consortium (2005). Growing by Degrees Online Education in the United States 2005, by I. E. Allen & J. Seaman. http://onlinelearningconsortium.org/publications/survey/growing_by_degrees_2005 (accessed 4 August 2014).

Tollan A. (2002). Land-use change and floods: what do we need most, research or management? *Water Science & Technology*, **45**(8), 183–190.

Tomboulian P., Schweitzer L., Mullin K., Wilson J. and Khiari, D. (2004). Materials used in drinking water distribution systems: contribution to taste-and-odor. *Water Science and Technology*, **49**(9), 219–226. www.documents.dgs.ca.gov/bsc/prpsd_chngs/documents/2007/pex/PEX%20EIR%20271%20Tomboulian,%20P.%20 2004.pdf (accessed 5 August 2014).

Tunay O., Kabdasli I., Arslan-Alaton I. and Olmez-Hanci T. (2010). Chemical Oxidation Applications for Industrial Wastewaters. IWA Publishing, UK.

Ujang Z. and Henze, M. (2003). Environmental biotechnology: advancement in water and wastewater application in the tropics. Selected Proceedings of the IWA International Conference Held in Kuala Lumpur, Malaysia, 9–10 December 2003, IWA Publishing, UK.

UNICEF/WHO Report (2012). Progress on Drinking Water and Sanitation 2012 Update. www.unicef.org/publications/index_69025.html (accessed 5 August 2014).

United States Environmental Protection Agency. Water: Private Wells Section Website. http://water.epa.gov/drink/info/well/basicinformation.cfm (accessed August 5, 2014).

United States Environmental Protection Agency (2002c). Wastewater Technology Fact Sheet: Sewers, Pressure. http://nepis.epa.gov/Exe/ZyPURL.cgi?Dockey=P10099Q2.txt (accessed 30 July 2014).

United States Environmental Protection Agency (2004). Safe Drinking Water Act Factsheet: Drinking Water Treatment. http://nepis.epa.gov/Exe/ZyPURL.cgi?Dockey=3000660S.txt (accessed 15 August 2014).

United States Environmental Protection Agency (2005a). Handbook for Managing Onsite and Clustered (Decentralized) Wastewater Treatment Systems. http://nepis.epa.gov/Exe/ZyPURL.cgi?Dockey=20017K2G.txt (accessed 2 August 2014).

United States Environmental Protection Agency (2005b). Manual for the Certification of Laboratories Analyzing Drinking Water – Criteria and Procedures Quality Assurance. 5th edn. http://nepis.epa.gov/Exe/ZyPURL.cgi?Dockey=30006MXP.txt (accessed 30 July 2014).

United States Environmental Protection Agency (2006a). Water System Operator Roles and Responsibilities: A Best Practices Guide. http://nepis.epa.gov/Exe/ZyPURL.cgi?Dockey=2000ZZBC.txt (accessed 11 August 2014).

United States Environmental Protection Agency (2006b). Emerging Technologies for Biosolids Management. http://nepis.epa.gov/Exe/ZyPURL.cgi?Dockey=P1006DGM.txt (accessed 28 July 2014).

United States Environmental Protection Agency (2013). Emerging Technologies for Wastewater Treatment and In-Plant Wet-Weather Management. http://nepis.epa.gov/Exe/ZyPURL.cgi?Dockey=P100ILDC.txt (accessed 2 August 2014).

United States Environmental Protection Agency, Office of Water (1999a). Guidance Manual for Compliance with the Interim Enhanced Surface Water Treatment Rule: Turbidity Provisions. http://nepis.epa.gov/Exe/ZyPURL.cgi?Dockey=200021OB.txt (accessed 28 July 2014).

United States Environmental Protection Agency, Office of Water (April 1999b). Alternative Disinfectants and Oxidants Guidance Manual. http://nepis.epa.gov/Exe/ZyPURL.cgi?Dockey=2000229L.txt (accessed 25 July 2014).

United States Environmental Protection Agency, Office of Water (September 1999c). Wastewater Technology Fact Sheet – Ozone Disinfection. http://nepis.epa.gov/Exe/ZyPURL.cgi?Dockey=200044E8.txt (accessed 25 July 2014).

United States Environmental Protection Agency, Office of Water (September 2000a). Wastewater Technology Fact Sheet – Dechlorination. http://nepis.epa.gov/Exe/ZyPURL.cgi?Dockey=P1001L40.txt (accessed 25 July 2014).

United States Environmental Protection Agency, Office of Water (2000b). Collection Systems Technology Fact Sheet: Sewers, Lift Station. http://nepis.epa.gov/Exe/ZyPURL.cgi?Dockey=901U0X00.txt (accessed 28 July 2014).

United States Environmental Protection Agency, Office of Water (2000c). Wastewater Technology Fact Sheet: Sewers, Force Main. http://nepis.epa.gov/Exe/ZyPURL.cgi?Dockey=P10099PU.txt (accessed 30 July 2014).

United States Environmental Protection Agency, Office of Water (2000d). Wastewater Technology Fact Sheet: Pipe Construction and Materials. http://nepis.epa.gov/Exe/ZyPURL.cgi?Dockey=P1000W83.txt (accessed 30 July 2014).

United States Environmental Protection Agency, Office of Water (2000e). Biosolids Technology Fact Sheet: Centrifuge Thickening and Dewatering. http://nepis.epa.gov/Exe/ZyPURL.cgi?Dockey=901U0S00.txt (accessed 6 August 2014).

United States Environmental Protection Agency, Office of Water (2002a). Collection Systems Technology Fact Sheet: Sewers, Conventional Gravity. http://nepis.epa.gov/Exe/ZyPURL.cgi?Dockey=P10053D9.txt (accessed 2 August 2014).

United States Environmental Protection Agency, Office of Water (2002b). (revised in 2008). Onsite Wastewater Treatment Systems Manual. http://nepis.epa.gov/Exe/ZyPURL.cgi?Dockey=30004GXI.txt (accessed 28 July 2014).

United States Environmental Protection Agency, Office of Water (2003a). Biosolids Technology Fact Sheet Gravity Thickening. http://nepis.epa.gov/Exe/ZyPURL.cgi?Dockey=P10053EB.txt (accessed 28 July 2014).

United States Environmental Protection Agency, Office of Water (2003b). Environmental Regulations and Technology: Control of Pathogens and Vector Attraction in Sewage Sludge Guide. http://nepis.epa.gov/Exe/ZyPURL.cgi?Dockey=40001LJV.txt (accessed 6 August 2014).

United States Fire Administration (USFA) (October 2008). Water Supply Systems and Evaluation Methods, Volume I: Water Supply System Concepts. www.usfa.fema.gov/fireservice/prevention_education/strategies/water_supply (accessed 28 July 2014).

United States Geological Survey (USGS). Groundwater Wells Section Website. http://water.usgs.gov/edu/earthgwwells.html (accessed 5 August 2014).

United States Geological Survey (USGS). Water Cycle Systems Section Website. http://water.usgs.gov/edu/watercycle.html (accessed 5 August 2014).

Válek P., Kohusová K., Salaj M., Kelbl M., Young T. and Vykydal Miroslav (2011). Implementing new regulatory mechanisms in Czech Republic contracts to meet EU requirements. *Water Utility Management International*, **6.2**, 11.

van Haandel A. C. and van der Lubbe J. G. M. (2012). Handbook of Biological Wastewater Treatment – Design and Optimization of Activated Sludge System. 2nd edn, IWA Publishing, UK.

Vanhooren H., Meirlaen J., Amerlinck Y., Claeys F., Vangheluwe H. and Vanrolleghem P. A. (2003). WEST: modelling biological wastewater treatment. *Journal of Hydroinformatics*, **5**, 27–50.

van Nieuwenhuijzen A. F., Havekes M., Reitsma B. A. and de Jong P. (2009). Wastewater treatment plant Amsterdam West: new, large, high-tech and sustainable. *Water Practice & Technology*, **4**(1), 1–8.

Waite M. (2010). *Sustainable Water Resources in the Built Environment*. IWA Publishing, UK.

Wallace S. (2006). Feasibility, design criteria, and O&M requirements for small scale constructed wetland wastewater treatment systems. http://www.iwapublishing.com/template.cfm?name=isbn1843397285.

Wang Wenyi, Zeng Weihua, Yao Bo & Wei Jing (2014). A simulation impact evaluation of social-economic development on water resource use. *Journal of Water Reuse and Desalination*, 4(3), 137–153.

Washington State Department of Ecology (WS DOE) (2010). Procedural Manual for the Environmental Laboratory Accreditation Program. https://fortress.wa.gov/ecy/publications/publications/1003048.pdf (accessed 25 July 2014).

Washington State Department of Transportation (WSDOT) (2007). Project Management Glossary of Terms. www.wsdot.wa.gov/publications/fulltext/ProjectMgmt/PMOG/PM_Glossary.pdf (accessed 24 July 2014).

Washington State Department of Transportation (WSDOT), Environmental and Engineering Programs, Strategic Analysis and Estimation Office (July 2009). Cost Estimating Manual for WSDOT Projects, Administrative Manual. http://www.wsdot.wa.gov/publications/manuals/fulltext/m3034/estimatingguidelines.pdf (accessed 30 July 2014).

Water Environment Research Foundation (WERF) (2009). Performance & Cost of Decentralized Unit Processes, Vacuum Sewer Systems: Factsheet C4. www.werf.org/c/DecentralizedCost/C4_Vacuum_Sewers.aspx (accessed May 2014).

Water Research Foundation (WRF) (2011). Assessing Seawater Intake Systems for Desalination Plants" prepared by E. D. Mackey, N. Pozos, W. James, T. Seacord, H. Hunt & D. L. Mayer. www.waterrf.org/Pages/Projects.aspx?PID=4080 (accessed 5 August 2014).

Water Reuse Association (2011). Overview of Desalination Plants Intake Alternatives White Paper. www.watereuse.org/sites/default/files/u8/Intake_White_Paper.pdf (accessed 5 August 2014).

WHO and IRC Water and Sanitation Centre (2003). Linking technology choice with operation and maintenance in the context of community water supply and sanitation – A Reference Document for Planners and Project Staff, prepared by F. Brikké & M. Bredero. www.who.int/water_sanitation_health/hygiene/om/wsh9241562153/en (accessed 5 August 2014).

World Health Organization (2008), Geneva. Prüss-Üstün A, Bos R, Gore F, Bartram J. Safer water, better health – costs, benefits and sustainability of interventions to protect and promote health. www.who.int/water_sanitation_health/publications/safer_water/en (accessed 5 August 2014).

Wuertz S., Bishop P. and Wilderer P. (2008). Biofilms in Wastewater Treatment: An Interdisciplinary Approach. IWA Publishing, UK.

Yan Z. S., Wang S. H., Kang X. K. and Ma, Y. (2010). Removal of organic compounds of real textile dyeing wastewater by pilot scale combination process of anoxic baffled reactor-hybrid membrane bioreactor. *Water Practice & Technology*, **55**(2), 1–8.

Yang Y., Lesage G., Barret M., Bernet N., Grasmick A., Hamelin J. and Heran M. (2013). New urban wastewater treatment with autotrophic membrane bioreactor at low chemical oxygen demand/N substrate ratio. *Water Science & Technology*, **69**(5), 960–965.

Examples of Employment Websites

Career Builder Job Board Website. www.careerbuilder.com (accessed 4 August 2014).

Glassdoor Job Board Website. www.glassdoor.com (accessed 4 August 2014).

Indeed Job Board Website. www.indeed.com (accessed 4 August 2014).

Monster Job Board Website. www.monster.com (accessed 4 August 2014).

Simply Hired Job Board Website. www.simplyhired.com (accessed 4 August 2014).

Appendix A

Glossary: Common water and wastewater terminologies

Acknowledgement

Glossary terms in the present section are used with the permission from the Office of Water Programs (OWP), California State University, Sacramento, California, USA and can be found online in the OWP glossary section at http://www.owp.csus.edu/glossary/index.php (accessed on July 24, 2014).

No.	Term	Meaning
1	Absorption	The taking in or soaking up of one substance into the body of another by molecular or chemical action (as tree roots absorb dissolved nutrients in the soil).
2	Activated sludge	Sludge particles produced in raw or settled wastewater (primary effluent) by the growth of organisms (including zoogleal bacteria) in aeration tanks in the presence of dissolved oxygen. The term activated comes from the fact that the particles are teeming with bacteria, fungi, and protozoa. Activated sludge is different from primary sludge in that the sludge particles contain many living organisms that can feed on the incoming wastewater.
3	Activated sludge process	A biological wastewater treatment process that speeds up the decomposition of wastes in the wastewater being treated. Activated sludge is added to wastewater and the mixture (mixed liquor) is aerated and agitated. After some time in the aeration tank, the activated sludge is allowed to settle out by sedimentation and is disposed of (wasted) or reused (returned to the aeration tank) as needed. The remaining wastewater then undergoes more treatment.
4	Aeration	The process of adding air to water. Air can be added to water by either passing air through water or passing water through air.
5	Aerobic bacteria	Bacteria that will live and reproduce only in an environment containing oxygen that is available for their respiration (breathing), namely atmospheric oxygen or oxygen dissolved in water. Oxygen combined chemically, such as in water molecules (H_2O), cannot be used for respiration by aerobic bacteria.

(*Continued*)

No.	Term	Meaning
6	Aerobic digestion	The breakdown of wastes by microorganisms in the presence of dissolved oxygen. This digestion process may be used to treat only waste activated sludge, or trickling filter sludge and primary (raw) sludge, or waste sludge from activated sludge treatment plants designed without primary settling. The sludge to be treated is placed in a large aerated tank where aerobic microorganisms decompose the organic matter in the sludge. This is an extension of the activated sludge process.
7	Alkalinity	The capacity of water or wastewater to neutralize acids. This capacity is caused by the water's content of carbonate, bicarbonate, hydroxide, and occasionally borate, silicate, and phosphate. Alkalinity is expressed in milligrams per litre of equivalent calcium carbonate. Alkalinity is not the same as pH because water does not have to be strongly basic (high pH) to have a high alkalinity. Alkalinity is a measure of how much acid must be added to a liquid to lower the pH to 4.5.
8	Anaerobic bacteria	Bacteria that live and reproduce in an environment containing no free or dissolved oxygen. Anaerobic bacteria obtain their oxygen supply by breaking down chemical compounds that contain oxygen, such as sulfate (SO_4^{2-}).
9	Anaerobic digestion	A treatment process in which wastewater solids and water (about 5 per cent solids, 95 per cent water) are placed in a large tank (the digester) where bacteria decompose the solids in the absence of dissolved oxygen. At least two general groups of bacteria act in balance: (1) saprophytic bacteria break down complex solids to volatile acids, the most common of which are acetic and propionic acids; and (2) methane fermenters break down the acids to methane, carbon dioxide, and water.
10	Anoxic	A condition in which the aquatic (water) environment does not contain dissolved oxygen (DO), which is called an oxygen deficient condition. Generally refers to an environment in which chemically bound oxygen, such as in nitrate, is present. The term is similar to anaerobic.
11	Aquifer	A natural, underground layer of porous, water-bearing materials (sand, gravel) usually capable of yielding a large amount or supply of water.
12	Attached growth process	Wastewater treatment processes in which the microorganisms and bacteria treating the wastes are attached to the media in the reactor. The wastes being treated flow over the media. Trickling filters and rotating biological contactors are attached growth reactors. These reactors can be used for BOD removal, nitrification, and denitrification.
13	Backwashing	The process of reversing the flow of water back through the filter media to remove the entrapped solids.
14	Benchmarking	A process an agency uses to gather and compare information about the productivity and performance of other similar agencies with its own information. The purpose of benchmarking is to identify best practices, set improvement targets, and measure progress.
15	Bioaugmentation	The addition of bacterial cultures to speed up the breakdown of grease and other organic materials. Bioaugmentation is used to clean sewers and, on a preventive basis, to remove deposits in sewers. This method is also used to prevent grease buildup in lift station wet wells.

No.	Term	Meaning
16	Biological process	A waste treatment process by which bacteria and other microorganisms break down complex organic materials into simple, nontoxic, more stable substances.
17	Biomass	A mass or clump of organic material consisting of living organisms feeding on the wastes in wastewater, dead organisms, and other debris.
18	Biosolids	A primarily organic solid product produced by wastewater treatment processes that can be beneficially recycled. The word biosolids is replacing the word sludge when referring to treated waste.
19	Buffer capacity	A measure of the capacity of a solution or liquid to neutralize acids or bases. This is a measure of the capacity of water or wastewater to offer a resistance to changes in pH.
20	Bypassing	The act of causing all or part of a flow to be diverted from its usual channels. In a wastewater treatment plant, overload flows should be bypassed into a holding pond for future treatment.
21	Calibration	A procedure that checks or adjusts an instrument's accuracy by comparison with a standard or reference.
22	Catch basin	A chamber or well used with storm or combined sewers as a means of removing grit that might otherwise enter and be deposited in sewers.
23	Centrifuge	A mechanical device that uses centrifugal or rotational forces to separate solids from liquids.
24	Clarification	Any process or combination of processes the main purpose of which is to reduce the concentration of suspended matter in a liquid.
25	Clear well	A reservoir for the storage of filtered water of sufficient capacity to prevent the need to vary the filtration rate with variations in demand. Also used to provide chlorine contact time for disinfection.
26	Coagulation	The clumping together of very fine particles into larger particles (floc) caused by the use of chemicals (coagulants). The chemicals neutralize the electrical charges of the fine particles, allowing them to come closer and form larger clumps. This clumping together makes it easier to separate the solids from the water by settling, skimming, draining, or filtering.
27	Combined sewer	A sewer designed to carry both sanitary wastewaters and stormwater or surface water runoff.
28	Comminutor	A device used to reduce the size of the solid materials in wastewater by shredding (comminution). The shredding action is like many scissors cutting to shreds all the large solids in the wastewater.
29	Corrosivity	An indication of the corrosiveness of a water. The corrosiveness of a water is described by the water's pH, alkalinity, hardness, temperature, total dissolved solids, dissolved oxygen concentration, and the Langelier Index.
30	Day tank	A tank used to store a chemical solution of known concentration for feed to a chemical feeder. A day tank usually stores sufficient chemical solution to properly treat the water being treated for at least one day. Also called an age tank.

(*Continued*)

No.	Term	Meaning
31	Debris	Any material in wastewater found floating, suspended, settled, or moving along the bottom of a sewer. This material may cause stoppages by getting hung up on roots or settling out in a sewer. Debris includes grit, paper, plastic, rubber, silt, and all materials except liquids.
32	Detention time	(1) The theoretical (calculated) time required for a given amount of water or wastewater to pass through a tank at a given rate of flow. (2) The time required to fill a tank at a given flow. (3) The actual time in hours, minutes, or seconds that a small amount of water is in a settling basin, flocculating basin, or rapid-mix chamber. In storage reservoirs, detention time is the length of time entering water will be held before being drafted for use (several weeks to years, several months being typical).
33	Dewater	(1) To remove or separate a portion of the water present in a sludge or slurry. To dry sludge so it can be handled and disposed of. (2) To remove or drain the water from a tank or a trench. A structure may be dewatered so that it can be inspected or repaired.
34	Digester	A tank in which sludge is placed to allow decomposition by microorganisms. Digestion may occur under anaerobic (more common) or aerobic conditions.
35	Disinfection by-product	A contaminant formed by the reaction of disinfection chemicals (such as chlorine) with other substances in the water being disinfected.
36	Dissolved oxygen	Molecular oxygen dissolved in water or wastewater, usually abbreviated DO.
37	Equalizing basin	A holding basin in which variations in flow and composition of a liquid are averaged. Such basins are used to provide a flow of reasonably uniform volume and composition to a treatment unit. Also called a balancing reservoir.
38	Eutrophication	The increase in the nutrient levels of a lake or other body of water; this usually causes an increase in the growth of aquatic animal and plant life.
39	Facultative bacteria	Facultative bacteria can use either dissolved oxygen or oxygen obtained from food materials such as sulfate or nitrate ions. In other words, facultative bacteria can live under aerobic, anoxic, or anaerobic conditions.
40	Filtration	The process of passing water through a porous bed of fine granular material to remove suspended matter from the water. The suspended matter is mainly particles of floc, soil, and debris; but it also includes living organisms such as algae, bacteria, viruses, and protozoa.
41	Flocculation	The gathering together of fine particles after coagulation to form larger particles by a process of gentle mixing. This clumping together makes it easier to separate the solids from the water by settling, skimming, draining, or filtering.
42	Fluoridation	The addition of a chemical to increase the concentration of fluoride ions in drinking water to a predetermined optimum limit to reduce the incidence (number) of dental caries (tooth decay) in children. Defluoridation is the removal of excess fluoride in drinking water to prevent the mottling (brown stains) of teeth.

No.	Term	Meaning
43	Food/microorganism (F/M) ratio	Food to microorganism ratio. A measure of food provided to bacteria in an aeration tank.
44	Free chlorine	Free chlorine is chlorine (Cl_2) in a liquid or gaseous form. Free chlorine combines with water to form hypochlorous (HOCl) and hydrochloric (HCl) acids. In wastewater, free chlorine usually combines with an amine (ammonia or nitrogen) or other organic compounds to form combined chlorine compounds.
45	Friction loss	The head, pressure, or energy (they are the same) lost by water flowing in a pipe or channel as a result of turbulence caused by the velocity of the flowing water and the roughness of the pipe, channel walls, or restrictions caused by fittings. Water flowing in a pipe loses head, pressure, or energy as a result of friction. Also called head loss.
46	Grab sample	A single sample of water collected at a particular time and place that represents the composition of the water only at that time and place.
47	Grinder pump	A small, submersible, centrifugal pump with an impeller, designed to grind solids into small pieces before they enter the collection system.
48	Grit chamber	A detention chamber or an enlargement of a collection line designed to reduce the velocity of flow of the liquid to permit the separation of mineral solids from organic solids by differential sedimentation.
49	Groundwater	Subsurface water in the saturation zone from which wells and springs are fed. In a strict sense the term applies only to water below the water table. Also called phreatic water and plerotic water.
50	Headworks	The facilities where wastewater enters a wastewater treatment plant. The headworks may consist of bar racks or bar screens, comminutors, a wet well, and pumps.
51	Hydraulic grade line	The surface or profile of water flowing in an open channel or a pipe flowing partially full. If a pipe is under pressure, the hydraulic grade line is that level water would rise to in a small, vertical tube connected to the pipe.
52	Incineration	The conversion of dewatered wastewater solids by combustion (burning) to ash, carbon dioxide, and water vapor.
53	Infiltration	The seepage of groundwater into a sewer system, including service connections. Seepage frequently occurs through defective or cracked pipes, pipe joints and connections, interceptor access risers and covers, or manhole walls.
54	Inflow	Water discharged into a sewer system and service connections from such sources as, but not limited to, roof leaders, cellars, yard and area drains, foundation drains, cooling water discharges, drains from springs and swampy areas, around manhole covers or through holes in the covers, cross-connections from storm and combined sewer systems, catch basins, stormwaters, surface runoff, street wash waters, or drainage. Inflow differs from infiltration in that it is a direct discharge into the sewer rather than a leak in the sewer itself.

(*Continued*)

No.	Term	Meaning
55	Ion exchange	A water or wastewater treatment process involving the reversible interchange (switching) of ions between the water being treated and the solid resin contained within an ion exchange unit. Undesirable ions are exchanged with acceptable ions on the resin or recoverable ions in the water being treated are exchanged with other acceptable ions on the resin.
56	Jar test	A laboratory procedure that simulates coagulation/flocculation with differing chemical doses. The purpose of the procedure is to estimate the minimum coagulant dose required to achieve certain water quality goals. Samples of water to be treated are placed in six jars. Various amounts of chemicals are added to each jar, stirred, and the settling of solids is observed. The lowest dose of chemicals that provides satisfactory settling is the dose used to treat the water.
57	Launders	(1) Sedimentation basin and filter discharge channels consisting of overflow weir plates (in sedimentation basins) and conveying troughs. (2) Sedimentation tank effluent troughs consisting of overflow weir plates.
58	Manhole	An opening in a sewer provided for the purpose of permitting operators or equipment to enter or leave a sewer. Sometimes called an access hole or a maintenance hole.
59	Mean cell residence time (MCRT)	An expression of the average time (days) that a microorganism will spend in the activated sludge process.
60	Mixed liquor suspended solids	The amount (mg/L) of suspended solids in the mixed liquor of an aeration tank.
61	Muffle furnace	A small oven capable of reaching temperatures up to 600°C (1,112°F). Muffle furnaces are used in laboratories for burning or incinerating samples to determine the amounts of volatile solids or fixed solids in samples of wastewater.
62	Nitrification	An aerobic process in which bacteria change the ammonia and organic nitrogen in water or wastewater into oxidized nitrogen (usually nitrate).
63	Normality	The number of gram-equivalent weights of solute in one litre of solution. The equivalent weight of any material is the weight that would react with or be produced by the reaction of 8.0 grams of oxygen or 1.0 gram of hydrogen. Normality is used for certain calculations of quantitative analysis.
64	O & M manual	Operation and maintenance manual. A manual that describes detailed procedures for operators to follow to operate and maintain a specific treatment plant and the equipment of that plant.
65	Ozonation	The application of ozone to water, wastewater, or air, generally for the purposes of disinfection or odor control.
66	Package treatment plant	A small wastewater treatment plant often fabricated at the manufacturer's factory, hauled to the site, and installed as one facility. The package may be either a small primary or a secondary wastewater treatment plant.
67	Pathogens (pathogenic organisms)	Organisms, including bacteria, viruses, protozoa, or internal parasites, capable of causing diseases (such as giardiasis, cryptosporidiosis, typhoid fever, cholera, or infectious hapatitis) in a host (such as a person). (There are many types of organisms that do not cause disease. These organisms are called nonpathogenic.)

No.	Term	Meaning
68	Pipe bedding	The prepared base or bottom of a trench or excavation on which a pipe is supported.
69	Point source	A discharge that comes out of the end of a pipe or other clearly identifiable conveyance. Examples of point source conveyances from which pollutants may be discharged include: ditches, channels, tunnels, conduits, wells, containers, rolling stock, concentrated animal feeding operations, landfill leachate collection systems, vessels, or other floating craft. A nonpoint source refers to runoff or a discharge from a field or similar source.
70	Potable water	Water that does not contain objectionable pollution, contamination, minerals, or infective agents and is considered satisfactory for drinking.
71	Pre-aeration	The addition of air at the initial stages of treatment to freshen the wastewater, remove gases, add oxygen, promote flotation of grease, and aid coagulation.
72	Precipitation	1. The total measurable supply of water received directly from clouds as rain, snow, hail, or sleet; usually expressed as depth in a day, month, or year, and designated as daily, monthly, or annual precipitation. 2. The process by which atmospheric moisture is discharged onto land or water surfaces. 3. The chemical transformation of a substance in solution into an insoluble form (precipitate).
73	Primary clarifier	A wastewater treatment device that consists of a rectangular or circular tank that allows those substances in wastewater that readily settle or float to be separated from the wastewater being treated.
74	Pump station	A wastewater pumping station that lifts the wastewater to a higher elevation when continuing the sewer at reasonable slopes would involve excessive depths of trench. Also, an installation of pumps that raise wastewater from areas too low to drain into available sewers.
75	Raw wastewater	Plant influent or wastewater before any treatment.
76	Re-aeration	The introduction of air through forced air diffusers into the lower layers of the reservoir. As the air bubbles form and rise through the water, oxygen from the air dissolves into the water and replenishes the dissolved oxygen. The rising bubbles also cause the lower waters to rise to the surface where oxygen from the atmosphere is transferred to the water. This is sometimes called surface re-aeration.
77	Receiving water	A stream, river, lake, ocean, or other surface or groundwaters into which treated or untreated wastewater is discharged.
78	Recirculation	The return of part of the effluent from a treatment process to the incoming flow.
79	Residual chlorine	The concentration of chlorine present in water after the chlorine demand has been satisfied. The concentration is expressed in terms of the total chlorine residual, which includes both the free and combined or chemically bound chlorine residuals. Also called chlorine residual.
80	Return activated sludge	Settled activated sludge that is collected in the secondary clarifier or the membrane basin and returned to the aeration basin to mix with incoming raw or primary settled wastewater.

(*Continued*)

No.	Term	Meaning
81	Runoff	That part of rain or other precipitation that runs off the surface of a drainage area and does not enter the soil or the sewer system as inflow.
82	Safe water	Water that does not contain harmful bacteria, or toxic materials or chemicals. Water may have taste and odor problems, color, and certain mineral problems and still be considered safe for drinking.
83	Sanitary sewer	A pipe or conduit (sewer) intended to carry wastewater or waterborne wastes from homes, businesses, and industries to the treatment works. Stormwater runoff or unpolluted water should be collected and transported in a separate system of pipes or conduits (storm sewers) to natural watercourses.
84	SCADA system	Supervisory Control And Data Acquisition system. A computer-monitored alarm, response, control, and data acquisition system used to monitor and adjust treatment processes and facilities.
85	Secondary treatment	A wastewater treatment process used to convert dissolved or suspended materials into a form more readily separated from the water being treated. Usually, the process follows primary treatment by sedimentation. The process commonly is a type of biological treatment followed by secondary clarifiers that allow the solids to settle out from the water being treated.
86	Sedimentation basin	A tank or basin in which water or wastewater is held for a period of time during which the heavier solids settle to the bottom and the lighter materials float to the surface. Also called settling tank or clarifier.
87	Seed sludge	In wastewater treatment, seed, seed culture, or seed sludge refer to a mass of sludge that contains populations of microorganisms. When a seed sludge is mixed with wastewater or sludge being treated, the process of biological decomposition takes place more rapidly.
88	Septic tank	A system sometimes used in which wastewater collection systems and treatment plants are not available. The system is a settling tank in which settled sludge and floatable scum are in intimate contact with the wastewater flowing through the tank and the organic solids are decomposed by anaerobic bacterial action. Used to treat wastewater and produce an effluent that is usually discharged by subsurface leaching. Also referred to as an interceptor; however, the preferred term is septic tank.
89	Sludge age	A measure of the length of time a particle of suspended solids has been retained in the activated sludge process.
90	Sludge volume index	A calculation that indicates the tendency of activated sludge solids (aerated solids) to thicken or to become concentrated during the sedimentation/thickening process. SVI is calculated in the following manner: (1) allow a mixed liquor sample from the aeration basin to settle for 30 minutes; (2) determine the suspended solids concentration for a sample of the same mixed liquor; (3) calculate SVI by dividing the measured (or observed) wet volume (mL/L) of the settled sludge by the dry weight concentration of MLSS in grams/L.
91	Stabilized waste	A waste that has been treated or decomposed to the extent that, if discharged or released, its rate and state of decomposition would be such that the waste would not cause a nuisance or odors in the receiving water.

No.	Term	Meaning
92	Tertiary treatment	Any process of water renovation that upgrades treated wastewater to meet specific reuse requirements. May include general cleanup of water or removal of specific parts of wastes insufficiently removed by conventional treatment processes. Typical processes include chemical treatment and pressure filtration. Also called advanced waste treatment.
93	Totalizer	A device or meter that continuously measures and sums a process rate variable in cumulative fashion over a given time period. For example, total flows displayed in gallons per minute, million gallons per day, cubic feet per second, or some other unit of volume per time period. Also called an integrator.
94	Trickling filter	A treatment process in which wastewater trickling over media enables the formation of slimes or biomass, which contain organisms that feed upon and remove wastes from the water being treated.
95	Turbidity	The cloudy appearance of water caused by the presence of suspended and colloidal matter. In the waterworks field, a turbidity measurement is used to indicate the clarity of water. Technically, turbidity is an optical property of the water based on the amount of light reflected by suspended particles. Turbidity cannot be directly equated to suspended solids because white particles reflect more light than dark-colored particles and many small particles will reflect more light than an equivalent large particle.
96	Variable frequency drive	A control system that allows the frequency of the current applied to a motor to be varied. The motor is connected to a low-frequency source while standing still; the frequency is then increased gradually until the motor and pump (or other driven machine) are operating at the desired speed.
97	Volatile solids	Those solids in water, wastewater, or other liquids that are lost on ignition of the dry solids at 550°C (1,022°F). Also called organic solids and volatile matter.
98	Water table	The upper surface of the zone of saturation of groundwater in an unconfined aquifer.
99	Wet well	A compartment or tank in which wastewater is collected. The suction pipe of a pump may be connected to the wet well or a submersible pump may be located in the wet well.
100	Yield	The quantity of water (expressed as a rate of flow—GPM, GPH, GPD, cu m/day, ML/day, or total quantity per year) that can be collected for a given use from surface or groundwater sources. The yield may vary with the use proposed, with the plan of development, and also with economic considerations.

Appendix B

Glossary: Construction management terminologies

Acknowledgement

Glossary terms in the present section are obtained from the following sources:

(1) *Project Management Glossary of Terms, 2007* with the permission from Washington State Department of Transportation (WSDOT), USA and can be found at website www.wsdot.wa.gov/publications/fulltext/ProjectMgmt/PMOG/PM_Glossary.pdf (accessed 24 July 2014).
(2) *Handbook of PE Pipe,* with the permission from Plastic Pipe Institute Inc. (PPI), USA and can be found at website http://plasticpipe.org/pdf/glossary.pdf (accessed 24 July 2014).
(3) *Partnership for Public Procurement Glossary Section,* with the permission from National Institute of Governmental Purchasing (NIGP) – The Institute for Public Procurement, USA and can be found at website http://www.globalpublicprocurement.org/Resources/Glossary (accessed 24 July 2014).

No.	Term	Meaning
1	Acceptance criteria(1)	Those criteria, including performance requirements and essential conditions, which must be met before project deliverables are accepted.
2	Addendum/addenda(3)	A written change, addition, alteration, correction or revision to a bid, proposal or contract document. Addendum/addenda may be issued following a pre-bid/pre-proposal conference or as a result of a specification or work scope change to the solicitation.
3	Agreement(1)	A legal document that binds two or more parties to specific and implied obligations (e.g., a contract).
4	Architectural and engineering (A & E) services(3)	Professional services within the scope of the practice of architecture and professional engineering, as defined by the jurisdiction, usually involving research, design, development, construction, alteration, or repair of real property.
5	Balance sheet(3)	A financial summary of the dollar amounts of a firm's assets, liabilities, and owner's equity accounts at the end of an accounting period.

(Continued)

No.	Term	Meaning
6	Bedding[(2)]	The earth or other material on which a pipe or conduit is supported.
7	Bid[(3)]	A tender, proposal or quotation submitted in response to a solicitation from a contracting authority.
8	Bid bond[(3)]	An insurance agreement, accompanied by a monetary commitment, by which a third party (the surety) accepts liability and guarantees that the bidder will not withdraw the bid. The bidder will furnish bonds in the required amount and if the contract is awarded to the bonded bidder, the bidder will accept the contract as bid, or else the surety will pay a specific amount.
9	Bidder's list[(3)]	A listing of names and addresses of vendors from whom bids, proposals or quotations can be solicited. The list is generally retained in a retrievable database.
10	Bid documentation[(3)]	A file containing all of the information and records relating to the bid, which may include all of the original bids received, specifications, insurance requirements, addenda, bonds, correspondence and all other relevant data that may be subject to audit and further review.
11	Bid opening[(3)]	The official process in which sealed bids are opened, usually in the presence of one or more witnesses, at the time and place specified in the invitation for bid. The amount of each bid is recorded and bids are made available for public inspection. It may be open to the public.
12	Boring[(2)]	An earth-drilling process used for installing conduits or pipelines, or obtaining soil samples for evaluation and testing.
13	Budget[(1)]	The approved estimate for the project or any work breakdown structure component or any schedule activity.
14	Change order[(1)]	A written document between the owner and the contractor signed by the owner and the contractor authorizing a change in the work or an adjustment in the contract sum or the contract time. A change order may be signed by the architect or engineer, provided they have written authority from the owner for such a procedure and that a copy of such written authority is furnished to the contractor upon request. The contract sum and the contract time may be changed only by a change order. A change order may be in the form of additional compensation or time, or less compensation or time (known as a deduction from the contract); the amount deducted from the contract sum by change order.
15	Change order request[(1)]	A written document issued by the owner requesting an adjustment to the contract sum or an extension of the contract time; generally issued by the architect or the owner's representative.
16	Checklist[(1)]	Items listed together for convenience of comparison, or to ensure the actions associated with them are managed appropriately and not forgotten. An example is a list of items to be inspected that is created during quality planning and applied during quality control.
17	Chemical resistance[(2)]	Ability to render service in the transport of a specific chemical for a useful period of time at a specific concentration and temperature.
18	Closure[(1)]	The process of finalizing all activities across all of the project process groups to formally close the project or phase.

No.	Term	Meaning
19	Composite pipe[(2)]	Pipe consisting of two or more different materials arranged with specific functional purpose to serve as pipe.
20	Contingency allowance[(1)]	As a result of risk analysis, money or time may be set aside as contingency, which may be used in the event of risks occurring. Contingency allowance provides for variations, which may occur in the expected values of elements of cost or schedule, but not scope or quality.
21	Contract[(1)]	A contract is a mutually binding agreement, which obligates the seller to provide the specified product or service or result, and obligates the buyer to pay for it.
22	Contract closure[(1)]	The process of completing and settling the contract, including resolution of any open items, and closing each contract.
23	Contract documents[(1)]	A term used to represent all executed agreements between the owner and contractor; any general, supplementary, or other contract conditions; the drawings and specifications; all bidding documents, less bidding information, plus pre-award addenda issued prior to execution of the contract and post-award change orders; and any other items specifically stipulated as being included in the contract documents, which collectively form the contract between the contractor and the owner.
24	Customer[(1)]	The person or organization that will use the project's product or service or result.
25	Deliverable[(1)]	Any unique and verifiable product, result, or capability to perform a service that must be produced to complete a process, phase, or project. Often used more narrowly in reference to an external deliverable, which is a deliverable subject to approval by the project sponsor or customer.
26	Design-bid-build[(3)]	The traditional delivery method for construction projects in which design and construction are sequential and contracted for separately with two contracts.
27	Design-build contracting[(1)]	A contract structure in which both design and construction responsibility are vested in a single contractor.
28	Erosion[(2)]	Deterioration of a surface by the abrasive action of moving fluids. This is accelerated by the presence of solid particles or gas bubbles in suspension. When deterioration is further increased by corrosion, the term 'corrosion–erosion' is often used.
29	Engineer[(3)]	An individual, partnership, or corporation that performs professional engineering services for the agency as an independent contractor. A registered, professional engineer is an individual who works to develop economical and safe solutions to practical problems by applying mathematical and scientific knowledge while considering technical constraints. Engineers design materials, structures, machines and systems while considering the limitations imposed by practicality, safety and cost while adhering to local and national building codes. Engineers are required to have errors and omissions insurance.

(*Continued*)

No.	Term	Meaning
30	Estimate[(1)]	A quantitative assessment of the likely amount or outcome. Usually applied to project costs, resources, effort, and durations and is usually preceded by a modifier (i.e., preliminary, conceptual, feasibility, order-of-magnitude, definitive). It should always include some indication of accuracy (e.g., ±x per cent).
31	Final acceptance[(1)]	The action of the owner accepting the work from the contractor when the owner deems the work completed in accordance with the contract requirements. The owner when making the final payment to the contractor confirms final acceptance.
32	Final inspection[(1)]	A final site review of the project by the contractor, owner, or owner's authorized representative prior to issuing the final certificate for payment.
33	Final payment[(1)]	The last payment from the owner to the contractor of the entire unpaid balance of the contract sum, as adjusted by any approved change orders.
34	Fixed fee[(1)]	A set contract amount for all labor, materials, equipment and services; and contractor's overhead and profit for all work being performed for a specific scope of work.
35	Headwall[(2)]	A wall (of any material) at the end of a culvert or, drain to serve one or more of the following purposes: protect fill from scour or undermining; increase hydraulic efficiency, divert direction of flow, and serve as a retaining wall.
36	Hydraulic cleaning[(2)]	Techniques and methods used to clean sewer lines with water, for example, water pumped in the form of a high velocity spray and water flowing by gravity or head pressure. Devices include high velocity jet cleaners, cleaning balls, and hinged disc cleaners.
37	Inspector[(2)]	(1) The owner's on-site representative responsible for inspection and acceptance, approval, or rejection of work performed as set forth in the specifications. (2) An authorized representative of the engineer assigned to observe the work and report his findings to the engineer.
38	Invitation for bid[(3)]	All documents used to solicit competitive or multi-step sealed bids. Also known as invitation to bid (ITB).
39	Life cycle cost[(3)]	The total cost of ownership over the life span of the asset. A procurement technique that takes into account operating, maintenance, the time value of money, disposal and other associated costs of ownership as well as the residual value of the item.
40	Lump sum[(3)]	An aggregate or lot price which may represent the total price for a group of items in place of or in addition to unit prices for each individual item. The total price of a group of items which is priced as a whole for bidding purposes.
41	Mechanical cleaning[(2)]	Methods used to clean sewer lines of debris mechanically with devices such as rodding machines, bucket machines, winch-pulled brushes, and so on.
42	Milestone[(1)]	A significant point or event in the project.

No.	Term	Meaning
43	Notice of award[(3)]	A written notification from the public entity to the successful bidder, or offeror stating that there is an award of a contract in accordance with a bid or proposal previously submitted, and that effective with receipt the vendor or contractor shall proceed with performance; allows work to start while contract is printed and readied for distribution.
44	Open end contract[(3)]	A contract which sets forth the general provisions of supplies and services that may be delivered or performed within a given period of time, but in which quantity and/or duration is not specified. The quantity and delivery are specified with the placement of orders.
45	Output[(1)]	A product, result, or service generated by a process. May be an input to a successor process.
46	Per cent complete[(1)]	An estimate, expressed as a percentage, of the amount of work that has been completed on an activity or a work breakdown structure component.
47	Pipeline rehabilitation[(2)]	The *in situ* repair of an existing pipeline, which has become corroded or abraded, by insert renewal of a liner which rehabilitates the bore of the pipeline but does not contribute significantly to increased pressure capability or increased structural strength, yet does improve flow efficiency/hydraulics.
48	Planning processes[(1)]	Those processes performed to define and mature the project scope, develop the project management plan, and identify and schedule the project activities that occur within the project.
49	Pre-bid/pre-proposal conference[(3)]	A meeting held by the buyer with potential bidders/offerors, prior to the opening of the solicitation for the purpose of answering questions, clarifying any ambiguities and responding to general issues in order to establish a common basis for understanding all of the requirements of the solicitation. May result in the issuance of an addendum to all potential providers. In certain situations, a mandatory conference may be advisable.
50	Pre-design phase[(1)]	The phase prior to the start of design where feasibility studies are done and conceptual project cost estimates are prepared.
51	Prime contractor[(1)]	Any contractor having a contract directly with the owner. Usually the main (general) contractor for a specific project.
52	Procedure[(1)]	A series of steps followed in a regular definitive order to accomplish something.
53	Procurement documents[(1)]	Those documents utilized in bid and proposal activities, which include buyer's invitation for bid, invitation for negotiations, request for information, request for quotation, request for proposal, and seller's responses.
54	Progress meeting[(1)]	A meeting dedicated essentially to contractor progress during the construction phase.
55	Progress schedule[(1)]	A line diagram showing proposed and actual starting and completion times of the respective project activities.

(*Continued*)

No.	Term	Meaning
56	Project calendar[(1)]	A calendar of working days or shifts that establishes those dates on which schedule activities are worked, and nonworking days that determine those dates on which schedule activities are idle. Typically defines holidays, weekends, and shift hours.
57	Project cost[(1)]	All costs for a specific project, including costs for land, professionals, construction, furnishings, fixtures, equipment, financing, and any other project-related costs.
58	Project management[(1)]	The application of knowledge, skills, tools, and techniques to project activities to meet the project requirements.
59	Project schedule[(1)]	The planned dates for performing schedule activities and the planned dates for meeting schedule milestones.
60	Project team[(1)]	All the project team members, including the project management team, the project manager and, for some projects, the project sponsor.
61	Purchase order[(3)]	A purchaser's written document to a vendor formalizing all the terms and conditions of a proposed transaction, such as a description of the requested items, delivery schedule, terms of payment, and transportation.
62	Records management[(1)]	The practice of identifying, classifying, archiving, preserving, and destroying records.
63	Request for information (RFI)[(1)]	A type of procurement document whereby the buyer requests a potential seller to provide various pieces of information related to a product or service or seller capability.
64	Request for proposal (RFP)[(1)]	A type of procurement document used to request proposals from prospective sellers of products or services.
65	Request for qualification (RFQ)[(3)]	A document which is issued by a procurement entity to obtain statements of the qualifications of potential development teams or individuals (i.e., consultants) to gauge potential competition in the marketplace, prior to issuing the solicitation.
66	Retainage[(3)]	A specified amount or percentage of the progress payment due usually under a construction contract. Upon completion of all contract requirements, retained amounts must be paid promptly.
67	Retaining wall[(2)]	A wall for sustaining the pressure of earth or filling deposited behind it.
68	Rip rap[(2)]	Rough stone of various large sizes placed compactly or irregularly to prevent scour by water or debris.
69	Risk[(1)]	An uncertain event or condition that, if it occurs, has a positive or negative effect on a project's objectives.
70	Risk analysis[(1)]	An examination of risk areas or events to assess the probable consequences for each event (or combination of events in the analysis), and determine possible options for avoidance.
71	Scheduling[(1)]	The process of converting a general or outline plan for a project into a time-based schedule based on the available resources and time constraints.
72	Scope[(1)]	The sum of the products, services, and results to be provided as a project.

No.	Term	Meaning
73	Scope of work[(1)]	Defines the work to be done in detail, the materials to be used and the exact nature of the work to be done.
74	Seepage[(2)]	Water escaping through or emerging from the ground along a rather extensive line or surface, as contrasted with a spring, the water of which emerges from a single spot.
75	Specification[(1)]	A document that specifies, in a complete, precise, verifiable manner, the requirements, design, behavior, or other characteristics of a system, component, product, result, or service and, often, the procedures for determining whether these provisions have been satisfied. Examples are: requirement specification, design specification, product specification, and test specification.
76	Stakeholder[(1)]	Those with a particularly significant interest in the project's outcome, including those providing funding or right of way for the project and property owners who are affected by the project. Stakeholders are unique for each project.
77	Start date[(1)]	A point in time associated with a schedule activity's start, usually qualified by one of the following: actual, planned, estimated, scheduled, early, late, target, baseline, or current.
78	Subcontractor[(3)]	Any person or business entity employed to perform part of a contractual obligation under the control of the principal contractor. Any supplier, distributor, vendor, or firm that furnishes supplies or services to a prime contractor or another subcontractor.
79	Substantial completion[(1)]	The stage in the progress of the work when the work, or designated portion of the work, is sufficiently complete in accordance with the contract documents so that the owner can occupy or utilize the work for its intended use.
80	Team building[(1)]	The process of influencing a group of diverse individuals, each with their own goals, needs, and perspectives, to work together effectively for the good of the project such that their team will accomplish more than the sum of their individual efforts could otherwise achieve.
81	Technical proposal[(3)]	A response to a solicitation which describes in detail what an offeror proposes to furnish and the method of delivery. May be part of a two-step response contained within an offer, the second part being the price proposal.
82	Tender[(3)]	A bidding process for goods, services or construction that is open to all qualified bidders and where the sealed bids are opened in public for scrutiny and are chosen on the basis of compliance with the bid requirements and lowest price.
83	Validation[(1)]	The technique of evaluating a component or product during or at the end of a phase or project to ensure it complies with the specified requirements.
84	Value engineering[(3)]	A technique by which contractors may (1) voluntarily suggest methods for performing more economically and may share in any resulting savings or (2) be required to establish a program or identify and submit methods for performing more economically.

(*Continued*)

No.	Term	Meaning
85	Vendor[(3)]	A supplier/seller of goods and services. A reference to a provider of product or service.
86	War room[(1)]	A room used for project conferences and planning, often displaying charts of cost, schedule status, and other key project data.
87	Warranty[(3)]	A promise made by a seller to a buyer that is legally enforceable. The promise may be expressed or implied and is legally binding.
88	Watershed[(2)]	Region or area contributing to the supply of a stream or lake; drainage area, drainage basin, catchment area.
89	Work authorization[(1)]	A permission and direction, typically written, to begin work on a specific schedule activity or work package or control account. It is a method for sanctioning project work to ensure that the work is done by the identified organization, at the right time, and in the proper sequence.
90	Work order[(1)]	A written order, signed by the owner or his representative, of a contractual status requiring performance by the contractor without negotiation of any sort.
91	Work plan[(1)]	A comprehensive, realistic, and deliverable plan to accomplish the team mission and deliver the project. It includes initiate and align and plan the work elements, including a schedule and a budget.
92	Working pressure[(2)]	The maximum anticipated, sustained operating pressure applied to the pipe exclusive of transient pressures.

Appendix C

Glossary: Laboratory operations terminologies

Acknowledgement

Glossary terms in the present section are obtained from the following sources:

(1) *Quality Manual,* August 2013, with the permission from Laboratory and Environmental Assessment Division, State of Oregon Department of Environmental Quality (DEQ), USA and can be found online at http://www.deq.state.or.us/lab/techrpts/docs/DEQ91LAB0006QMP.pdf (accessed 25 July 2014).

(2) *Procedural Manual for the Environmental Laboratory Accreditation Program,* September 2010, with the permission from Washington State Department of Ecology, USA and can be found online at https://fortress.wa.gov/ecy/publications/publications/1003048.pdf (accessed 25 July 2014).

No.	Term	Meaning
1	Accuracy(1)	The degree of agreement between an observed value and an accepted reference value. Accuracy includes a combination of random error (precision) and systematic error (bias) components which are due to sampling and analytical operations; a data quality indicator.
2	Analyst(1)	The designated individual who performs the 'hands-on' analytical methods and associated techniques and who is the one responsible for applying required laboratory practices and other pertinent quality controls to meet the required level of quality.
3	Analytical data(2)	The qualitative or quantitative results from a chemical, physical, microbiological, toxicological, radiochemical, or other scientific determination.
4	Audit(1)	A systematic evaluation to determine the conformance to quantitative and qualitative specifications of some operational function or activity.
5	Batch(2)	A set of samples analyzed together without interruption. Results are usually calculated from the same calibration curve or factor.

(*Continued*)

No.	Term	Meaning
6	Blank sample(2)	A synthetic sample, free of the analyte(s) of interest. For example, in water analysis, pure water is used for the blank. In chemical analysis, a blank is used to estimate the analytical response to all factors other than the analyte in the sample. *Field blanks* are used to obtain information on contamination introduced during sample collection, transport, or storage. *Method blanks* are used to reveal contamination introduced by the laboratory.
7	Calibration standard(2)	Solution of a known analyte concentration, used in the calibration procedure to determine the relationship between concentration and analytical response.
8	Chain of custody form(1)	Record that documents the possession of the samples from the time of collection to receipt in the laboratory. This record generally includes: the number and types of containers; the mode of collection; collector; time of collection; preservation; and requested analyses.
9	Data integrity(1)	The result of the processes that together assure valid data of known and documented quality.
10	Document control(1)	The act of ensuring that documents (and revision thereto) are proposed, reviewed for accuracy, approved for release by authorized personnel, distributed properly, and controlled to ensure use of the correct version at the location where the prescribed activity is performed.
11	Environmental laboratory(2)	A facility in a specific geographic location, owned or managed by a single entity where scientific determinations are performed on samples taken from the environment, including drinking water samples.
12	Field measurement(1)	The determination of physical, biological, or radiological properties, or chemical constituents that are measured on-site, close in time and space to the matrices being sampled/measured, following accepted test methods. This testing is performed in the field outside of a fixed-laboratory or outside of an enclosed structure that meets the requirements of a mobile laboratory.
13	Holding time(2)	The allowed time from when a sample was taken or extracted until it must be analyzed. For composited samples, the holding time starts when the last composite aliquot is collected.
14	Matrix(2)	The substance from which a sample is collected, such as groundwater, ambient water, wastewater, air, solid, semisolid (such as tissue), or chemical compounds (such as oil).
15	Method(2)	A formalized group of procedures and techniques for performing an activity (e.g., sampling, chemical analysis, data analysis), systematically presented in the order in which they are to be executed.
16	Precision(2)	A measure of the variability in the results of replicate measurements caused by random error. Also referred to as imprecision. Precision is usually measured as standard deviation, relative standard deviation (RSD), or relative percentage difference (RPD).
17	Quality assurance (QA)(2)	A set of activities designed to establish and document the reliability and usability of measurement data.
18	QA manual(2)	A QA manual documents policies, organization, objectives, and specific QC and QA activities. Volume and scope of QA manuals vary with complexity of the laboratory mission.

No.	Term	Meaning
19	Quality control (QC)[2]	The routine application of statistically based procedures to assess the accuracy of measurement data.
20	Quality system[1]	A structured and documented management system describing the policies, objectives, principles, organizational authority, responsibilities, accountability, and implementation plan of an organization for ensuring quality in its work processes, products (items), and services. The quality system provides the framework for planning, implementing, and assessing work performed by the organization and for carrying out required QA and QC.
21	Random error[2]	Variability in the results of replicate measurements. Random error is so named because the size and magnitude of the difference between replicate results vary at random and not in any systematic way.
22	Reference materials[1]	Material or substance one or more of whose property values are sufficiently homogeneous and well established to be used for the calibration of an apparatus, the assessment of a measurement method, or for assigning values to materials.
23	Reference standard[1]	Standard used for the calibration of working measurement standards in a given organization or at a given location.
24	Sampling[1]	Activity related to obtaining a representative sample of the object of conformity assessment, according to a procedure.
25	Standard deviation[2]	A statistic that describes the random variability in results of repeated measurements.
26	Standard operating procedures (SOPs)[1]	A written document which details the method of an operation, analysis or action whose techniques and procedures are thoroughly prescribed and which is accepted as the method for performing certain routine or repetitive tasks.
27	Systematic errors[2]	Errors that cause measurement results to be consistently greater or smaller than the true value. Usually bias can be considered to be equivalent to systematic error.

Appendix D

Common conversion factors and basic formulae

Acknowledgement

Conversion factors and basic formulae in the present section are obtained from the following sources:

(1) *Wastewater Treatment Mathematical Formulas,* 2nd Edition, 2011, with the permission from Fleming Training Center, Division of Water Resources, Tennessee Department of Environment and Conservation, USA and can be found at website http://www.tn.gov/environment/water/fleming_math-books.shtml (accessed 24 July 2014).

(2) *Applied Wastewater Math Formula Sheet and Conversion Factors,* with the permission from Division of Drinking and Ground Waters, Ohio Environmental Protection Agency, USA and can be found at website http://www.epa.ohio.gov/portals/28/documents/opcert/WWformulas.pdf (accessed 24 July 2014).

(3) *Wastewater Treatment Conversion Factors and Formulas,* with the permission from Kentucky Division of Compliance Assistance, Energy and Environment Cabinet, Department of Environmental Protection, USA and can be found at website http://dca.ky.gov/OPCert/certification/Documents/WWConversionTable[1].pdf (accessed 24 July 2014).

Conversion factors

No.	Parameter	Formula
1	Length	1 inch = 2.54 centimetres 1 foot (ft) = 12 inches (in) 1 meter (m) = 3.28 feet 1 yard (yd) = 3 feet (ft) 1 mile (mi) = 5280 feet (ft) 1 mile = 1760 yds

(Continued)

Conversion factors (*Continued*).

No.	Parameter	Formula
2	Area	1 ft^2 = 144 in^2 1 Acre = 43,560 ft^2 1 mi^2 = 640 acres 1 acre = 0.405 hectare (Ha) 1 Hectare = 2.47 acres
3	Volume	1 liter (L) = 1000 milliliters (mL) 1 US gallon = 3.785 liters (L) 1 liter = 0.2642 US gallons (gal) 1 quart = 946 mL 1 liter = 1.06 quarts 1 ft^3 = 7.48 US gallons 1 ft^3 = 28.32 liters 1 yd^3 = 27 ft^3 1 US gallon = 231 in^3 1 US gallon = 0.1337 ft^3 1 million US gallon (MG) = 1,000,000 US gallons 1 acre/foot (ac-ft) = 43,560 ft^3 1 acre-foot (ac-ft) = 325,851 US gallons (gal)
4	Weight and mass	1 gram (gm) = 1000 milligrams (mg) 1 kilogram (kg) = 1000 grams (g) 1 kilogram = 2.2 pounds (lbs) 1 pound (lb) = 453.6 grams 1 ounce = 28.35 gms 1 pound = 7000 grains 1 US ton = 2000 lbs 1 mL of water = 1 gm 1 US gallon of water = 8.34 lbs 1 ft^3 of water = 62.4 lbs 1% = 10,000 milligrams per liter (mg/L) 1 mg/L = 0.0584 grains per US gallon 1 grain per US gallon = 17.118 mg/L
5	Pressure and head	1 ft head of water = 0.433 lbs/in^2 (psi) 1 lbs/in^2 (psi) = 2.31 ft head of water 1 psi = 2.036 inches of mercury 1 inch of mercury = 0.4912 psi
6	Time	1 minute (min) = 60 seconds (sec) 1 hour (hr) = 60 minutes 1 day = 24 hours 1 day = 1440 minutes
7	Flow	1 million US gallons per day (MGD) = 694 gal/min 1 MGD = 11.57 gal/sec 1 MGD = 1.55 ft^3/sec 1 ft^3/sec = 449 gal/min 1 ft^3/sec = 0.6463 MGD 1 gal/min = 0.00144 MGD

Conversion factors (*Continued*).

No.	Parameter	Formula
8	Temperature	°C = (°F − 32°) ÷ 1.8 °F = (1.8 × °C) + 32°
9	Power	1 kWh = 3413 Btu's 1 hp = 0.746 kilowatts 1 hp = 42.44 Btu/Min 1 hp = 746 watts 1 hp = 33,000 ft lbs/min 1 hp = 550 ft lbs/sec

Area formulae

No.	Parameter	Formula
1	Rectangle	Area (ft^2) = (Length, ft) × (Width, ft)
2	Circle	Area (ft^2) = 0.785 × (Diameter, ft)^2
3	Triangle	Area (ft^2) = 0.5 × (Base, ft) × (Height, ft)

Volume formulae

No.	Parameter	Formula
1	Rectangle	Volume (ft^3) = (Length, ft) × (Width, ft) × (Height, ft)
2	Cylinder	Volume (ft^3) = 0.785 × (Diameter, ft)^2 × (Depth or Length, ft)
3	Cone	$\text{Volume (ft}^3) = \frac{0.785 \times (\text{Diameter, ft})^2 \times (\text{Depth, ft})}{3}$
4	Sphere	Volume (ft^3) = 0.5236 × (Diameter)^3
5	Pyramid	$\text{Volume (ft}^3) = \frac{(\text{Height, ft}) \times (\text{Base Area, ft}^2)}{3}$

Flow related formulae

No.	Parameter	Formula
1	Velocity	$\text{Velocity (ft/sec)} = \frac{\text{Distance (ft)}}{\text{Time (sec)}}$
2	Flow	Flow (ft^3/sec) = (Area, ft^2) × (Velocity, ft/sec)
3	Flow (channel)	Flow (ft^3/sec) = (Width, ft) × (Depth, ft) × (Velocity, ft/sec)
4	Flow (pipe)	Flow (ft^3/sec) = 0.785 × (Diameter, ft)^2 × (Velocity, ft/sec)
5	Average daily flow (ADF)	$\text{ADF(MGD)} = \frac{\text{Sum of all daily flows, MGD}}{\text{Number of daily flows}}$
6	Per capita daily flow	$\text{Daily flow(gal/day/capita)} = \frac{\text{Water used, gal/day}}{\text{Number of people served}}$

(*Continued*)

Power formulae

No.	Parameter	Formula
1	Power	$\text{Power(watts)} = (\text{Volts}) \times (\text{Amperes})$
2	Amps, single-phase	$\text{Amps} = \dfrac{(746\ \text{watts/HP})(\text{Horsepower})}{(\text{Volts})(\text{Efficiency, \%, as decimal})(\text{Power Factor})}$
3	Amps, three-phase	$\text{Amps} = \dfrac{(746\ \text{watts/HP})(\text{Horsepower})}{(1.732)(\text{Volts})(\text{Efficiency, \%, as decimal})(\text{Power Factor})}$
4	Horsepower	$\text{Horsepower} = \dfrac{(\text{Volts})(\text{Amperes})}{(746\ \text{watts/HP})}$
5	Kilowatt, single-phase	$\text{Kilowatts} = \dfrac{(\text{Volts})(\text{Amperes})(\text{Power Factor})}{(1000\ \text{watts/kilowatt})}$
6	Kilowatt, three-phase	$\text{Kilowatts} = \dfrac{(1.732)(\text{Volts})(\text{Amperes})(\text{Power Factor})}{(1000\ \text{watts/kilowatt})}$
7	Power factor	$\text{Power Factor} = \text{watts} / (\text{Volts})(\text{Amperes})$

Pumps formulae

No.	Parameter	Formula
1	Pumping rate	$\text{Pumping rate (gal/min)} = \dfrac{\text{Volume (gal)}}{\text{Time (min)}}$
2	Time to fill	$\text{Time to fill (min)} = \dfrac{\text{Tank Volume (gal)}}{\text{Flow Rate (gal/min)}}$
3	Water HP	$\text{Water HP} = \dfrac{(\text{Flow, gal/min})(\text{Head, ft})}{3960}$
4	Brake HP	$\text{Breake HP} = \dfrac{(\text{Flow, gal/min})(\text{Head, ft})}{(3960)(\text{Pump Efficiency, \%, as decimal})}$
5	Brake HP	$\text{Breake HP} = \dfrac{(\text{Water Horsepower})}{(\text{Pump Efficiency, \%, as decimal})}$
6	Motor HP	$\text{Motor HP} = \dfrac{(\text{Brake Horsepower})}{(\text{Motor Efficiency, \%, as decimal})}$
7	Pump efficiency	$\text{Pump efficiency, \%} = \dfrac{(\text{Water Horsepower})}{(\text{Brake Horsepower})} \times 100\%$
8	Motor efficiency	$\text{Motor efficiency, \%} = \dfrac{(\text{Brake Horsepower})}{(\text{Motor Horsepower})} \times 100\%$

Pumps formulae (*Continued*).

No.	Parameter	Formula
9	Overall efficiency	$\text{Overall efficiency, \%} = \dfrac{(\text{Water Horsepower})}{(\text{Motor Horsepower})} \times 100\%$
10	System head	Static Head, ft = Suction Lift, ft + Discharge Head, ft Static Head, ft = Discharge Head, ft – Suction Head, ft Total Dynamic Head, ft = Static Head, ft + Friction Losses, ft
11	Pumping cost	$\text{Cost, \$/hr} = (\text{Motor horsepower})\left(0.746\,\dfrac{\text{kW}}{\text{HP}}\right)\left(\text{Cost, }\dfrac{\$}{\text{kW} - \text{hr}}\right)$

Appendix E

Common drinking water system formulae

Acknowledgement

Conversion factors and basic formulae are obtained from the following source:

Water Treatment Mathematical Formulas, 6th Edition, 2011, with the permission from Fleming Training Center, Division of Water Resources, Tennessee Department of Environment and Conservation, USA and can be found at website http://www.tn.gov/environment/water/fleming_math-books.shtml (accessed 24 July 2014).

No.	Parameter	Formula
1	Water wells	$\text{Well yield, gal/min} = \dfrac{\text{Volume, gal}}{\text{Time, min}}$
		Drawdown, ft = Pumping water level, ft – Static water level, ft
		$\text{Specific capacity, gal/min/ft} = \dfrac{\text{Well yield, gal/min}}{\text{Drawdown, ft}}$
2	Coagulation/flocculation	$\text{Detention time, min} = \dfrac{(\text{Volume, gal})(1440\ \text{min/day})}{\text{Flow, gal/day}}$
3	Sedimentation	$\text{Detention time, hrs} = \dfrac{(\text{Volume, gal})(24\ \text{hrs/day})}{\text{Flow, gal/day}}$
		$\text{Surface Overflow Rate (SOR), gal/day/ft}^2 = \dfrac{\text{Flow, gal/day}}{\text{Surface Area, ft}^2}$
		$\text{Weir Overflow Rate (WOR), gal/day/ft} = \dfrac{\text{Flow, gal/day}}{\text{Weir Length, ft}}$
		Circular weir length, ft = (3.14)(Weir Diameter, ft) Rectangular weir length, ft = 2(Weir Length, ft) + 2(Weir Width, ft)

(*Continued*)

No.	Parameter	Formula
4	Filtration	Filter production rate, gal/min = (Filtration Rate, gal/min/ft^2) × (Filter Area, ft^2) Backwash water volume, gal $= \left(\text{Backwash Rate, } \frac{\text{gal}}{\text{min, ft}^2}\right)$(Backwash time, min)(Filter Area, ft^2) Backwash water, $\% = \frac{\text{Backwash water, gal}}{\text{Water filtered, gal}} \times 100\%$
5	Chlorination	Cl_2 Demand, mg/L = Cl_2 Dose, mg/L – Cl_2 Residual, mg/L Total chlorine residual, mg/L = Combined residual, mg/L + Free residual, mg/L $Cl_2 \text{ Feed Rate, } \frac{\text{lbs}}{\text{day}} = \frac{\text{(Dosage, mg/L)(Flow, MGD)(8.34 lbs/gal)}}{\text{Chemical purity, \%, as decimal}}$
6	Average feed rate (Dry feeder)	$\text{Average feed rate, grams/min} = \frac{\text{Total sample mass, grams}}{\text{Sample collection time, min}}$
7	Average feed rate (Liquid feeder)	$\text{Average feed rate, mL/min} = \frac{\text{Total sample volume, mL}}{\text{Sample collection time, min}}$
8	Chemical feed systems	Chemical feed, lbs = (Dose, mg/L) (Volume, MG) (8.34 lbs/gal) Chemical feed rate, lbs/day = (Dose, mg/L) (Flow, MGD) (8.34 lbs/gal)

Appendix F

Common wastewater system formulae

Acknowledgement

Conversion factors and basic formulae are obtained from the following source:

Wastewater Treatment Mathematical Formulas, 2nd Edition, 2011, with the permission from Fleming Training Center, Division of Water Resources, Tennessee Department of Environment and Conservation, USA and can be found at website http://www.tn.gov/environment/water/fleming_math-books.shtml (accessed 24 July 2014).

No.	Parameter	Formula
1	Primary treatment	$\text{Screenings removed, ft}^3\text{/day} = \dfrac{\text{Screenings, ft}^3}{\text{Day}}$
		$\text{Screenings removed, ft}^3\text{/MG} = \dfrac{\text{Screenings, ft}^3}{\text{Flow, MG}}$
		$\text{Screening pit capacity, days} = \dfrac{\text{Screening pit volume, ft}^3}{\text{Screenings removed, ft}^3\text{/day}}$
		$\text{Grit removed, ft}^3\text{/MG} = \dfrac{\text{Grit volume, ft}^3}{\text{Flow, MG}}$
2	Sedimentation	$\text{Detention time, hrs} = \dfrac{\text{(Volume of basin, gal)(24 hrs/day)}}{\text{Flow, gal/day}}$
		$\text{Weir overflow rate, gal/day/ft} = \dfrac{\text{Flow, gal/day}}{\text{Total weir length, ft}}$
		Circular basin weir length, ft = (3.14)(Diameter, ft) Rectangular basin weir length, ft = (2)(Length, ft) + (2)(Width, ft)

(Continued)

No.	Parameter	Formula
		$\text{Surface overflow rate, gal/day/ft}^2 = \dfrac{\text{Flow, gal/day}}{\text{Surface area, ft}^2}$
		$\text{Solids loading rate, lbs/day/ft}^2 = \dfrac{\text{Solids applied, lbs/day}}{\text{Surface area, ft}^2}$
		BOD Removed, lbs/day = (BOD removed, mg/L)(Flow, MGD)(8.34 lbs/gal)
		$\text{Efficiency, \%} = \dfrac{\text{In} - \text{Out}}{\text{In}} \times 100\%$
3	Activated sludge	BOD loading, lbs/day = (BOD, mg/L)(Flow, MGD)(8.34 lbs/gal)
		COD loading, lbs/day = (COD, mg/L)(Flow, MGD)(8.34 lbs/gal)
		MLSS, lbs = (MLSS, mg/L)(Aerator volume, MG)(8.34 lbs/gal)
		MLVSS, lbs = (MLSS, lbs)(Volatile solids, % expressed as decimal)
		$\text{F/M Ratio} = \dfrac{\text{BOD or COD, lbs/day}}{\text{Mixed liquor volatile suspended solids, lbs}}$
		$\text{F/M Ratio} = \dfrac{\text{BOD or COD, lbs/day}}{\text{(MLSS, lbs)(Volatile solids, \% expressed as decimal)}}$
		$\text{Return activated sludge flow, \%} = \dfrac{\text{Return sludge flow, MGD}}{\text{Influent flow, MGD}} \times 100\%$
		$\text{Return sludge concentration, mg/L} = \dfrac{1{,}000{,}000}{\text{SVI}}$
		MCRT = Mean Cell Residence Time
		$\text{MCRT, days} = \dfrac{\text{Suspended solids in system, lbs}}{\text{Suspended solids leaving system, lbs/day}}$
		$\text{MCRT, days} = \dfrac{\text{Suspended solids in system, lbs}}{\text{Waste activated sludge SS, lbs/day} + \text{Secondary effluent SS, lbs/day}}$
		$\text{Minutes to waste sludge} = \dfrac{\text{MLSS, gal to waste}}{\text{Waste sludge pump capacity, gal/min}}$
4	Trickling filter	$\text{Hydraulic loading rate, gal/day/ft}^2 = \dfrac{\text{Flow rate, gal/day}}{\text{Surface area, ft}^2}$
		$\text{Organic (BOD) loading, lbs/day/1000 ft}^3 = \dfrac{\text{BOD Applied, lbs/day}}{\text{Volume of media, 1000 ft}^3}$
		$\text{Volume of media, 1000 ft}^3 = \dfrac{(0.785)(\text{Trickling filter diameter, ft})^2(\text{Media depth, ft})}{1000}$
5	Sludge digestion	$\text{Solids added, lbs} = \left(\text{Flow}, \dfrac{\text{gal}}{\text{day}}\right)\left(8.34\,\dfrac{\text{lbs}}{\text{gal}}\right)(\text{Total Solids Conc., \% as decimal})$
		$\text{Seed sludge concentration, \%} = \dfrac{\text{Seed sludge volume, gal}}{\text{Total digester capacity, gal}} \times 100\%$

No.	Parameter	Formula
		VS = Volatile Solids and TS = Total Solids
		$\text{VS Concentration, \%} = \dfrac{\text{Volatile solids feed rate, lbs/day}}{\text{Total solids feed rate, lbs/day}} \times 100\%$
		$\text{VS feed rate, } \dfrac{\text{lbs}}{\text{day}} = \left(\text{TS feed rate, } \dfrac{\text{lbs}}{\text{day}}\right)(\text{VS Conc., \% as decimal})$
		$\text{Volatile acids/Alkalinity ratio} = \dfrac{\text{Volatile acids concentration, mg/L}}{\text{Total alkalinity, mg/L as } CaCO_3}$
		$\text{VS Reduction, \%} = \dfrac{\text{Volatile solids reduced, lbs}}{\text{Total volatile solids entering, lbs}} \times 100\%$
		$\text{VS Destroyed, lbs/day/ft}^3 = \dfrac{\text{Volatile solids destroyed, lbs/day}}{\text{Digester volume, ft}^3}$
		$\text{Gas produced, ft}^3\text{/lb VS Destroyed} = \dfrac{\text{Gas produced, ft}^3}{\text{Volatile solids destroyed, lbs}}$
		$\text{Digestion time, days} = \dfrac{\text{Digester volume, gal}}{\text{Sludge flow rate, gal/day}}$
6	Sludge production and thickening	$\text{Solids concentration, \%} = \dfrac{\text{Solids loading rate, lbs/day}}{\text{Sludge production rate, lbs/day}} \times 100\%$
		$\text{Hydraulic loading rate, gal/day/ft}^2 = \dfrac{\text{Total flow, gal/day}}{\text{Area, ft}^2}$
		$\text{Solids loading rate, lbs/day/ft}^2 = \dfrac{\text{Solids loading rate, lbs/day}}{\text{Area, ft}^2}$
		$\text{Sludge detention time, days} = \dfrac{\text{Sludge blanket volume, gal}}{\text{Sludge pumping rate from thickener, gal/day}}$
7	Sludge dewatering (Filter press)	$\text{Solids loading rate, lbs/hr/ft}^2 = \dfrac{\text{Sludge feed rate, lbs/hr}}{\text{Plate area, ft}^2}$
8	Sludge dewatering (Belt press)	$\text{Hydraulic loading rate, gal/min/ft} = \dfrac{\text{Flow, gal/min}}{\text{Belt width, ft}}$
		$\text{Sludge feed rate, lbs/hr} = \dfrac{\text{Sludge to be dewatered, lbs/day}}{\text{Operating time, hr/day}}$
9	Sludge dewatering (Vacuum filter)	$\text{Filter loading, lbs/hr/ft}^2 = \dfrac{\text{Solids to filter, lbs/hr}}{\text{Surface area, ft}^2}$

Appendix G

Common laboratory analysis formulae

Acknowledgement

Conversion factors and basic formulae are obtained from the following source:

Wastewater Treatment Mathematical Formulas, 2nd Edition, 2011, with the permission from Fleming Training Center, Division of Water Resources, Tennessee Department of Environment and Conservation, USA and can be found at website http://www.tn.gov/environment/water/fleming_math-books.shtml (accessed 24 July 2014).

No.	Parameter	Formula
1	Solution preparation	% Strength(by weight) $= \dfrac{\text{Weight of chemical, lbs}}{(\text{Weight of water, lbs}) + (\text{Weight of chemical, lbs})} \times 100\%$ $(\text{Concentration}_1)(\text{Volume}_1) = (\text{Concentration}_2)(\text{Volume}_2)$ $(\text{Normality}_1)(\text{Volume}_1) = (\text{Normality}_2)(\text{Volume}_2)$
2	Biochemical Oxygen Demand (BOD_5)	$BOD_5\text{, mg/L} = \dfrac{(D_1 - D_2)}{P}$ $BOD_5\text{, mg/L} = \dfrac{(D_1 - D_2) - (B_1 - B_2)f}{P}$ D_1 = Initial dissolved oxygen concentration in sample, mg/L D_2 = Final dissolved oxygen concentration in sample, mg/L B_1 = Initial dissolved oxygen concentration in seed control, mg/L B_2 = Final dissolved oxygen concentration in seed control, mg/L P = Sample concentration, % (expressed as a decimal) f = Seed in sample (%)/Seed in seed control (%)

(Continued)

No.	Parameter	Formula
3	Alkalinity	$\text{Total alkalinity}, \frac{\text{mg}}{\text{L}} \text{ as } CaCO_3 = \frac{(B)(N)(50{,}000)}{\text{Sample volume, mL}}$ B = Volume of titrant used, mL N = Normality of sulfuric acid
4	Sludge volume index (SVI)	$\text{SVI, mL/g} = \frac{(\text{Wet settled sludge volume, mL})(1000)}{\text{Dried sludge solids mass, mg}}$ $\text{SVI, mL/g} = \frac{(\text{Settled sludge volume, mL/L})(1000)}{\text{mixed liquor suspended solids, mg/L}}$
5	Sludge density index (SDI)	$\text{SDI} = \frac{100}{\text{SVI}}$
6	Suspended solids (SS)	$\text{SS, mg/L} = \frac{(A-B)(1{,}000{,}000)}{\text{Sample volume, mL}}$ A = Final weight of pan, filter, and residue in grams B = Weight of prepared filter and pan in grams
7	Total solids (TS)	$\text{TS, mg/L} = \frac{(A-B)(1{,}000{,}000)}{\text{Sample volume, mL}}$ A = Weight of dish and dried solids in grams B = Weight of dish in grams
8	Volatile solids (VS)	$\text{VS, mg/L} = \frac{(A-B)(1{,}000{,}000)}{\text{Sample volume, mL}}$ A = Weight of dish and dried solids in grams B = Weight of dish and ash in grams
9	Fixed solids (FS)	$\text{FS, mg/L} = \frac{(A-B)(1{,}000{,}000)}{\text{Sample volume, mL}}$ A = Weight of dish and ash in grams B = Weight of dish in grams
10	Volatile solids (%VS)	$\%\ \text{VS} = \frac{\text{Volatile solids, mg/L}}{\text{Total solids, mg/L}} \times 100\%$
11	Bacteriological colonies	$\text{Mean} = \frac{\text{Sum of items or values}}{\text{Number of items or values}}$ $\text{Geometric mean} = \sqrt[n]{(X_1)(X_2)(X_3)\ldots\ldots(X_n)}$ Bacteriological colonies/100 mL $= \frac{(\text{Number of colonies counted})(100)}{\text{Sample volume filtered, mL}}$

Appendix H

Listing: Water treatment system components

Following is a partial list of common components of the water treatment system.

Sr. No.	Equipment	Common process components
1	Rapid or flash mixing tank	• Chemicals tank, mixing tank, spill containment • Foundation, supports, skid mounts • Mechanical mixer, support bridge, tank baffles • Valves, fittings, strainer, drain • Chemical feed pumps, motors • Process piping, chemical injection port • Manway, access openings • Process instrumentation and controls • Ladder, maintenance platform, grating • Spare parts, chemicals, maintenance tools
2	Inline mixers (static/dynamic)	• Inline mixing unit • Inlet and outlet piping • Chemicals injection port, maintenance tools
3	Flocculation basin or tank	• Flocculation tank or basin • Tank foundation, walls • Water inlet and distribution piping • Weirs, orifices, baffles, diffuser walls • Low speed mixers, paddles, turbines, shafts • Outlet piping, valves, fittings • Drive units, process instrumentation and controls

(*Continued*)

Sr. No.	Equipment	Common process components
		• Access walkways, handrails, grating, ladders
		• Spare parts, maintenance tools
4	Sedimentation basin	• Settling tank or basin
		• Inlet and outlet piping
		• Distributors, weirs, launders, troughs
		• Tube settlers, parallel plates, baffles
		• Sediment settling zone, raking device
		• Sludge removal piping
		• Access walkways, handrails, grating, ladders
		• Valves, fittings, gates
		• Spare parts, maintenance tools
5	Clarifier tank	• Clarification tank
		• Inlet piping, stilling well
		• Skimmer, scum trough, draw-off piping
		• Tube settlers, parallel plates
		• Launder, weirs, piping, valves and fittings
		• Sludge rake arms, scrapper blades, concentrator
		• Drive assembly, shaft, motor, controls
		• Access bridge, walkways, handrails, grating
		• Spare parts, maintenance tools
6	Filtration system	• Filtration tank or unit, media bed, support plates
		• Unit foundation, supports, skid mounts
		• Water inlet and distribution piping
		• Backwash piping, wash trough
		• Outlet piping, overflow weirs, underdrain
		• Valves, control valves, drain, fittings
		• Process instrumentation and controls
		• Manway access openings, sight glass window
		• Ladder, rungs, grating, maintenance platform
		• Spare parts, filter media, maintenance tools
7	Ion exchange, adsorption units	• Ion exchange or adsorption unit
		• Unit foundation, supports, skid mounts
		• Media bed, media support plates
		• Water inlet pipe, distributor, jet breaker
		• Treated effluent water piping, underdrain
		• Regenerant inlet and outlet piping
		• Regeneration chemicals storage and feed system
		• Valves, control valves, fittings, flow meter
		• Sump, drain, sight glass window, manways
		• Access platform, ladder, gratings

Sr. No.	Equipment	Common process components
		• Process instrumentation and controls • Spare parts, media, maintenance tools
8	UV disinfection process	• Ultraviolet disinfection unit, channel, piping • Lamps, sleeves, wiper, drain, valves, fittings • Process instrumentation and controls • Spare UV bulbs, modules, maintenance tools
9	Chemical disinfection process	• Chemical storage and feed system • Disinfection contact tank • Tank foundation, walls, support structure • Inlet and outlet piping • Flow distribution structures, baffles • Valves, fittings, chemical injection port • Process instrumentation and controls • Spare parts, chemicals, maintenance tools
10	Clear well storage tank	• Clear well tank, foundation, walls, supports • Baffles, chemical injection ports • Mixers, air piping, compressor, controls • Spare parts, chemicals, feed pumps
11	Process chemicals feed systems	• Chemicals storage and spill containment • Mixing tank, foundation, supports, skid structures • Ladder, maintenance platform • Mechanical mixer, support bridge, tank baffles • Valves, fittings, floats, switches, injection port • Chemical feed pumps, motors • Manway, openings, access platform • Flow meter, process instrumentation and controls • Spare parts, chemicals, maintenance tools
12	Water plant pumping systems	• Pump stations units, pump room, well • Access manholes, hatches, vent • Ladder, rungs, maintenance area • Pumps, base, guide rail, hoist • Access platform, handrail, grating • Suction piping, discharge force mains • Valve vault, valves, fittings, flow meter • Pump motors, instrumentation and controls • Water level sensors, floats, lights, alarms • Spare parts, maintenance tools
13	Electronic flow meters	• Flow meter units, support structures • Inlet and outlet pipes

(*Continued*)

Sr. No.	Equipment	Common process components
		• Meter vault, valves, fittings • Transmitter, display unit, recorder • Spare parts, maintenance tools
14	Open channel flow meters	• Flow measurement weirs, flumes, channel • Inlet and outlet sections, piping, valves, fittings • Water level sensor, flow measurement unit • Spare parts, maintenance tools
15	Aeration blower systems	• Aeration blower units, supports, skid mount • Air inlet piping, inlet filter • Inlet and discharge silencers • Vibration isolation pads • Motor, slide base, drive guard, enclosure • Variable frequency drives (VFDs) • Pressure and vacuum relief valves • Isolation valves, check valves • Process instrumentation and controls • Spare parts, maintenance tools

BIBLIOGRAPHY

Alaska Department of Environmental Conservation, Division of Water (2013) Introduction to Small Water Systems Course Manual. http://dec.alaska.gov/water/opcert/IntroSmallCorrOnlineCourses.html (accessed 28 July 2014).

Arregui F., Cabrera E. Jr. and Cobacho R. (2006). Integrated Water Meter Management. IWA Publishing, UK.

Bratby J. (2006). Coagulation and Flocculation in Water and Wastewater Treatment. 2nd edn, IWA Publishing, UK.

Bell K. Y. and Abel S. (2011). Optimization of WWTP aeration process upgrades for energy efficiency. *Water Practice and Technology*, **6**(2). http://www.apg-neuros.com/Uploads/Documents/7%20-%20Aeration%20Optimization%20-%20IWA%20Publishings%20-%202011.pdf (accessed 28 July 2014).

Cao Z. and Zha F. (2006). Improving feed water recovery and reducing backwash waste in membrane filtration process. *Water Practice and Technology*, **1**(4).

Deininger A., Günthert F. W. and Wilderer P. A. (1996). The influence of currents on circular secondary clarifier performance and design. *Water Science and Technology*, **34**(3–4), 405–412.

Ghomi M. R., Sohrabnejad M. and Ovissipour M. R. (2009). An experimental study of nozzle diameters, aeration depths and angles on standard aeration efficiency (SAE) in a venturi aerator. *Water Practice and Technology*, **4**(3), 1–8.

Haarhoff J. and Joubert H. (1997). Determination of aggregation and breakup constants during flocculation. *Water Science and Technology*, **36**(4), 33–40.

Ignjatović L. (1991). Efficient (tube) settling tanks. *Water Science and Technology*, **24**(7), 223–228.

Nassehi V. and Das D. B. (2007). Computational Methods in the Management of Hydro-Environmental Systems. IWA Publishing, UK.

Park N. S., Kim S. S., Lee Y. J. and Wang C. K. (2013). Effects of longitudinal baffles on particles settling in a sedimentation basin. *Water Science and Technology*, **69**(6), 1212–1218.

Parker D., Butler R., Finger R., Fisher R., Fox W., Kido W., Merrill S., Newman G., Pope R., Slapper J. and Wahlberg E. (1996). Design and operations experience with flocculator-clarifiers in large plants. *Water Science and Technology*, **33**(12), 163–170.

Phu T. D., Huey D. P., Chung D. L. and Hurdle R. L. (2010). Treatment and disposal of water treatment plant residuals for Ho Chi Minh City, Vietnam. *Water Practice and Technology*, **5**(4), 1–12.

Safe Drinking Water Foundation, Canada (2009) Conventional Water Treatment: Coagulation and Filtration Factsheet. http://www.safewater.org/resources/fact-sheets.html (accessed 30 July 2014).

Saleh A. M. and Hamoda M. F. (1999). Upgrading of secondary clarifiers by inclined plate settlers. *Water Science and Technology*, **40**(7), 141–150.

Skovmose K. C. and Eriksen M. (2010). Supervision of pumps and their operating conditions in sewage pumping stations. *Water Practice and Technology*, **5**(2), 1–11.

State of Connecticut, Department of Public Health Drinking Water Section (2008). Liquid Chemical Feed System Design and Installation Guidelines. http://www.ct.gov/dph/lib/dph/drinking_water/pdf/Liquid_Chemical_Feed_System_Design_and_Construction_Guidelines.pdf (accessed 28 July 2014).

United States Environmental Protection Agency, Office of Water (1999). Guidance Manual for Compliance with the Interim Enhanced Surface Water Treatment Rule: Turbidity Provisions. http://nepis.epa.gov/Exe/ZyPURL.cgi?Dockey=200021OB.txt (accessed 28 July 2014).

United States Environmental Protection Agency, Office of Water (2000). Collection Systems Technology Fact Sheet: Sewers, Lift Station. http://nepis.epa.gov/Exe/ZyPURL.cgi?Dockey=901U0X00.txt (accessed 28 July 2014).

United States Environmental Protection Agency, Office of Water (2003). Biosolids Technology Fact Sheet Gravity Thickening. http://nepis.epa.gov/Exe/ZyPURL.cgi?Dockey=P10053EB.txt (accessed 28 July 2014).

United States Environmental Protection Agency, Office of Water (2009). Drinking Water Treatment Guidance Document. http://nepis.epa.gov/Exe/ZyPURL.cgi?Dockey=3000660S.txt (accessed 28 July 2014).

United States Fire Administration (2008) Water Supply Systems and Evaluation Methods, Volume I: Water Supply System Concepts. http://www.usfa.fema.gov/fireservice/prevention_education/strategies/water_supply (accessed 28 July 2014).

van Haandel A. C. and van der Lubbe J. G. M. (2012). Handbook of Biological Wastewater Treatment – Design and Optimization of Activated Sludge System. 2nd edn, IWA Publishing, UK.

van Nieuwenhuijzen A. and van der Graaf J. (2011). Handbook on Particle Separation Processes. IWA Publishing, UK.

van Staden S. J. and Haarhoff J. (2011). How to measure specific deposit washout and backwash efficiency of granular filters. *Water Practice and Technology*, **6**(1).

Wagner M. R. and Pöpel H. J. (1998). Oxygen transfer and aeration efficiency – influence of diffuser submergence, diffuser density, and blower type. *Water Science and Technology*, **38**(3), 1–6.

Wang S. and Garcia R. (2013). Development of a self-rechargeable digital water flow meter. *Journal of Hydroinformatics*, **15**(3), 888–896.

Appendix I

Listing: Wastewater treatment system components

Following is a partial list of common components of the wastewater treatment system.

No.	Equipment	Common process components
1	Influent pumping station	• Pump stations units, dry well and wet well • Access manholes, hatches, vent, bird screen • Ladder, rungs, maintenance area • Pumps, base, guide rail, hoist • Access platform, handrail, grating • Suction piping, discharge force mains • Valve vault, valves, fittings, flow meter • Pump motors, instrumentation and controls • Water level sensors, floats, lights, alarms • Spare parts, maintenance tools
2	Influent bar and fine screens	• Screening units, screen channel • Bypass channel, piping, valves, fittings • Discharge chute, screen collection dumpster • Conveyor system, drive unit, motor • Rake for debris removal, accessories • Screen backwashing system, wiper, washer • Ladder, rungs, grating, access platform, handrails • Process instrumentation and controls • Weather protection, insulation, ventilation • Spare parts, maintenance tools
3	Wastewater grinders	• Inline and channel grinders units • Motors, supports, couplings • Speed reducers, process instrumentation and controls

(*Continued*)

No.	Equipment	Common process components
		• Slide rails, cutters and spacers • Spare parts, maintenance tools
4	Grit removal system	• Grit removal unit, supporting structure, channels • Access platform, walkway, grating, handrails • Water inlet and outlet piping, bypass • Blowers, air piping, header, distribution, diffusers • Grit pump, motor, shaft • Grit discharge piping, trough, grit classifier • Grease skimmer, scum discharge piping • Grit unit baffles • Spare parts, maintenance tools
5	Dissolved air flotation (DAF)	• DAF unit, walkway bridge, grating, handrails • Water inlet piping, distribution structure • Outlet piping, trough, overflow, weirs, drain • Motor, drive unit, gear box, shaft • Air compressor, distribution piping, diffusers • Valves, control valves, fittings • Scum scraper, skimmer, collection box, piping • Rake arm, sludge removal mechanism, piping • Spare parts, maintenance tools
6	Clarifier tank	• Clarification tank • Inlet piping, stilling well • Skimmer, scum trough, draw-off piping • Tube settlers, parallel plates • Launder, weirs, piping, valves and fittings • Sludge concentrator section, rake arms • Scrapper blades, squeegees • Drive assembly, shaft, motor, controls • Access bridge, walkways, handrails, grating • Spare parts, maintenance tools
7	Biological aerobic treatment	• Aeration tank, foundation, walls, supports • Access bridge, walkway, grating, handrails • Wastewater inlet and outlet piping • Splitter boxes, baffles, troughs, weirs, decanter • Blowers, air distribution piping, diffusers • Aerators, mixers, motors, supports • Valves, control valves, fittings, samplers • Sludge recycling, sludge wasting pumps, motors • Process instrumentation and controls • Spare parts, maintenance tools

No.	Equipment	Common process components
8	Anoxic tank	• Anoxic tank, foundation, walls, supports • Access walkway, grating, handrails • Wastewater inlet piping, distribution • Baffles, troughs, outlet piping • Mixers, motors, supports • Valves, control valves, fittings • Process instrumentation and controls • Spare parts
9	Membrane bioreactors (MBR)	• MBR process tank, foundation, walls, supports • Access walkway, grating, handrails • Wastewater inlet piping, distribution structure • MBR modules, supporting panels, cleaning system • Blowers, air distribution piping, diffusers • Valves, control valves, fittings • Waste sludge pumps, motors, piping • Outlet piping, manifolds, permeate pumps • Process instrumentation and controls • Spare parts, maintenance tools
10	Constructed wetlands process	• Wetland units, foundation, base, walls, berms • Synthetic liner, wetland beds media, aquatic plants • Pumps, motor, controls, panels, instrumentation • Inlet zone, weir, distribution piping, inspection port • Outlet zone, collection piping, drain lines • Chemical feed, effluent pumps, recirculation • Spare parts, maintenance tools
11	UV disinfection process	• Ultraviolet disinfection unit, channel, piping • Lamps, sleeves, wiper, drain, valves, fittings • Process instrumentation and controls • Spare UV bulbs, modules, maintenance tools
12	Chemical disinfection process	• Chemical storage and feed system • Disinfection contact tank • Tank foundation, walls, support structure • Inlet and outlet piping • Flow distribution structures, baffles • Valves, fittings, chemical injection port • Process instrumentation and controls • Spare parts, chemicals, maintenance tools
13	Process chemicals feed systems	• Chemicals storage and spill containment • Mixing tank, foundation, supports, skid structures

(*Continued*)

No.	Equipment	Common process components
		• Ladder, maintenance platform • Mechanical mixer, support bridge, tank baffles • Valves, fittings, floats, switches, injection port • Chemical feed pumps, motors • Manway, openings, access platform • Flow meter, process instrumentation and controls • Spare Parts, Chemicals, Maintenance Tools
14	Electronic flow meters	• Flow meter units, support structures • Inlet and outlet pipes • Meter vault, valves, fittings • Transmitter, display unit, recorder • Spare parts, maintenance tools
15	Open channel flow meters	• Flow measurement weirs, flumes, channel • Inlet and outlet sections, piping, valves, fittings • Water level sensor, flow measurement unit • Spare parts, maintenance tools
16	Sludge thickening systems	• Gravity thickener, centrifuge, rotary drum units • Supports, access walkway, grating, handrails • Inlet and outlet piping, center pier • Launder, overflow weirs, baffles • Sludge rake arm, scraper, squeegees • Drive unit, motor, shaft • Chemical (polymer) feed system • Valves, fittings, process instrumentation and controls • Thickened sludge hopper, piping, sludge pumps • Spare parts, maintenance tools
17	Sludge stabilization systems	• Aerobic/anaerobic digester, compost, pasteurization • Inlet and outlet piping, sludge pumping system • Sludge mixer, motor, baffles • Digester cover, biogas collection, gas meter, sampler • Blowers, air distribution piping, valves, fittings • Sludge recycle system, sludge heater • Substrate effluent, sludge pumping, motor • Composting unit, piles, base, cover layers • Condensate trap, odor control, compost screening • Access manhole, vents, piping, fittings • Valves, fittings, process instrumentation and controls • Access walkway, grating, handrails, stairs • Spare parts, maintenance tools

No.	Equipment	Common process components
19	Sludge dewatering systems	• Belt/plate filter press, centrifuge, vacuum filter • Skid mount, supports, sludge feed, distribution • Drive unit, motor, polymer feed system, inline mixer • Inlet and discharge piping, valves, fittings, wash spray • Filter media, vacuum drum, vat, conveyor • Access walkway, grating, maintenance area • Process instrumentation and controls • Spare parts, maintenance tools
20	Sludge drying systems	• Drying unit (direct, indirect, flash) • Belts, drums, beds drying systems • Sludge intake and outlet structures • Air intake and exhaust, heat exchanger, solar • Reeds, filters, sludge turning system • Fittings, sensors, process instrumentation and controls • Spare parts, maintenance tools
21	Aeration blower systems	• Aeration blower units, supports, skid mount • Air inlet piping, inlet filter • Inlet and discharge silencers, vibration isolation pads • Motor, slide base, drive guard, enclosure • Variable frequency drives (VFDs) • Pressure and vacuum relief valves • Isolation valves, check valves • Process instrumentation and controls • Spare parts, maintenance tools

BIBLIOGRAPHY

Alaska Department of Environmental Conservation, Division of Water (2013) Introduction to Small Water Systems Course Manual. http://dec.alaska.gov/water/opcert/IntroSmallCorrOnlineCourses.html (accessed July 2014).

Alhakawati M. S., Cooper N. B., Al Asmi S., Abu El Soud M. and Abdulhussain H. (2014). Commissioning and operation experience of the largest flat sheet membrane bio-reactors in the world. *Water Practice and Technology*, **9**(2), 165–187.

Andreasen I. and Nielsen B. (1993). A comparative study of full-scale sludge dewatering equipment. *Water Science and Technology*, **28**(1), 37–45.

Arregui F., Cabrera E. Jr. and Cobacho R. (2006). Integrated Water Meter Management. IWA Publishing, UK.

Bell K. Y. and Abel S. (2011). Optimization of WWTP aeration process upgrades for energy efficiency. *Water Practice and Technology*, **6**(2). http://www.apg-neuros.com/Uploads/Documents/7%20-%20Aeration%20Optimization%20-%20IWA%20Publishings%20-%202011.pdf (accessed 28 July 2014).

Bouchy L., Senante E., Dauthuille P., Aupetitgendre M., Harry J. P., Venot S. and Rougé P. (2009). Odour creation potential of sludge during composting and drying. *Water Practice and Technology*, **4**(3), 1–9.

Brepols C. (2011). Operating Large Scale Membrane Bioreactors for Municipal Wastewater Treatment. IWA Publishing, UK.

Christian S., Grant S., McCarthy P., Wilson D. and Mills D. (2011). The first two years of full-scale anaerobic membrane bioreactor (AnMBR) operation treating high-strength industrial wastewater. *Water Practice and Technology*, **6**(2).

Deininger A., Günthert F. W. and Wilderer P. A. (1996). The influence of currents on circular secondary clarifier performance and design. *Water Science and Technology*, **34**(3–4), 405–412.

du Toit G. J. G., Ramphao M. C., Parco V., Wentzel M. C. and Ekama G. A. (2007). Design and performance of BNR activated sludge systems with flat sheet membranes for solid–liquid separation. *Water Science and Technology*, **56**(6), 105–113.

Edzwald J. K. (1995). Principles and applications of dissolved air flotation. *Water Science and Technology*, **31**(3–4), 1–23.

Engelhardt N. and Lindner W. (2006). Experiences with the world's largest municipal waste water treatment plant using membrane Technology. *Water Practice and Technology*, **1**(4).

European Commission (1999) Proceedings of the workshop on 'Problems around sludge', 18–19 November 1999, Stresa (Italy). http://ec.europa.eu/environment/waste/sludge/problems.htm (accessed July 2014).

Fabregat A., Bengoa C., Font J. and Stueber F. (2011). Reduction, Modification and Valorization of Sludge (Removals). IWA Publishing, UK.

Foladori P., Andreottola G. and Ziglio G. (2010). Sludge Reduction Technologies in Wastewater Treatment Plants. IWA Publishing, UK.

Frost R. C., Halliday J. and Dee A. S. (1993). Continuous consolidation of sludge in large scale gravity thickeners. *Water Science and Technology*, **28**(1), 77–86.

Ghomi M. R., Sohrabnejad M. and Ovissipour M. R. (2009). An experimental study of nozzle diameters, aeration depths and angles on standard aeration efficiency (SAE) in a venturi aerator. *Water Practice and Technology*, **4**(3), 1–8.

Ghosh S. (1991). Pilot-scale demonstration of two-phase anaerobic digestion of activated sludge. *Water Science and Technology*, **23**(7–9), 1179–1188.

Gruber Y., Farchill D. and Goldstein M. (1988). Design, testing and operation of grit chambers with hydraulic roll. *Water Science and Technology*, **20**(4–5), 237–248.

Gunes K., Tuncsiper B., Masi F., Ayaz S., Leszczynska D., Fındık Hecan N. and Ahmad H. (2011). Construction and maintenance cost analyzing of constructed wetland systems. *Water Practice and Technology*, **6**(3).

Irizar I., Alferes J., Larrea L. and Ayesa E. (2009). Design, implementation and experimental validation of a real-time control architecture for nitrogen removal in SBR systems. *Water Practice and Technology*, **4**(3), 1–8.

Kosobucki P., Chmarzyński A. and Buszewski B. (2000). Sewage sludge composting. *Polish Journal of Environmental Studies*, **9**(4), 243–248.

Larsen T., Udert K. and Lienert J. (2013). Source Separation and Decentralization for Wastewater Management. IWA Publishing, UK.

Libabher M. and Orozco-Jaramillo A. (2012). Sustainable Treatment and Reuse of Municipal Wastewater. IWA Publishing, UK.

Lundh M., Jönsson L. and Dahlquist J. (2001). The flow structure in the separation zone of a DAF pilot plant and the relation with bubble concentration. *Water Science and Technology*, **43**(8), 185–194.

Nassehi V. and Das D. B. (2007). Computational Methods in the Management of Hydro-Environmental Systems. IWA Publishing, UK.

Ohandja D.-G., Gove L., Edwards M., Callan J. and Voulvoulis N. (2010). Improving anaerobic digester performance: lessons learnt from digester clean-out operations. *Water Practice and Technology*, **5**(2), 1–9.

Oikonomidis I. and Marinos C. Solar sludge drying in Pafos wastewater treatment plant: operational experiences. *Water Practice and Technology*, **9**(1), 62–70.

Parker D., Butler R., Finger R. D, Fisher R., Fox W., Kido W., Merrill S., Newman G., Pope R., Slapper J. and Wahlberg E. (1996). Design and operations experience with flocculator-clarifiers in large plants. *Water Science and Technology*, **33**(12), 163–170.

Reverdy A. L., Dieudé-Fauvel E., Ferstler V. and Baudez J. C. (2013). Anaerobic digestion of sewage sludge: overview of the French situation. *Water Practice and Technology*, **8**(2), 180–189.

Saleh A. M. and Hamoda M. F. (1999). Upgrading of secondary clarifiers by inclined plate settlers. *Water Science and Technology*, **40**(7), 141–150.

Shilton A. (2005). Pond Treatment Technology. IWA Publishing, UK.

Skovmose K. C. and Eriksen M. (2010). Supervision of pumps and their operating conditions in sewage pumping stations. *Water Practice and Technology*, **5**(2), 1–11.

State of Connecticut, Department of Public Health Drinking Water Section (2008) Liquid Chemical Feed System Design and Installation Guidelines. http://www.ct.gov/dph/lib/dph/drinking_water/pdf/Liquid_Chemical_Feed_System_Design_and_Construction_Guidelines.pdf (accessed 28 July 2014).

Taia C. S., Snider-Nevina J., Dragasevicha J. and Kempsonb J. (2014). Five years operation of a decentralized membrane bioreactor package plant treating domestic wastewater. *Water Practice and Technology*, **9**(2).

United States Environmental Protection Agency, Office of Water (1999) Wastewater Technology Factsheet: Ultraviolet Disinfection. http://nepis.epa.gov/Exe/ZyPURL.cgi?Dockey=200044EE.txt (accessed 28 July 2014).

United States Environmental Protection Agency, Office of Water (2000) Collection Systems Technology Fact Sheet: Sewers, Lift Station. http://nepis.epa.gov/Exe/ZyPURL.cgi?Dockey=901U0X00.txt (accessed 28 July 2014).

United States Environmental Protection Agency, Office of Water (2000) Design Manual: Constructed Wetlands Treatment of Municipal Wastewaters. http://water.epa.gov/type/wetlands/restore/cwetlands.cfm (accessed 28 July 2014)

United States Environmental Protection Agency, Office of Water (2003) Biosolids Technology Fact Sheet Gravity Thickening. http://nepis.epa.gov/Exe/ZyPURL.cgi?Dockey=P10053EB.txt (accessed 28 July 2014).

United States Environmental Protection Agency, Office of Water (2006) Emerging Technologies for Biosolids Management. http://nepis.epa.gov/Exe/ZyPURL.cgi?Dockey=P1006DGM.txt (accessed 28 July 2014).

United States Environmental Protection Agency, Office of Water (2008) Onsite Wastewater Treatment Systems Manual. http://nepis.epa.gov/Exe/ZyPURL.cgi?Dockey=30004GXI.txt (accessed 28 July 2014).

van Bentem, A. G. N., Petri C. P. and Schyns P. F. T. (2007). Membrane Bioreactors: Operation and Results of a MBR Wastewater Treatment Plant. IWA Publishing, UK.

van Nieuwenhuijzen A. and van der Graaf J. (2011). Handbook on Particle Separation Processes. IWA Publishing, UK.

van Haandel A. C. and van der Lubbe J. G. M. (2012). Handbook of Biological Wastewater Treatment – Design and Optimization of Activated Sludge System. 2nd edn, IWA Publishing, UK.

Younga T., Muftugila M., Smoota S. and Peetersb J. (2012). MBR vs. CAS: capital and operating cost evaluation. *Water Practice and Technology*, **7**(4).

Appendix J

Listing: Chemicals used for treatment processes

Following is a partial list of chemicals commonly used in water and wastewater treatment processes.

No.	Chemical	Formula	Water or wastewater treatment
1	Activated alumina	Al_2O_3	Adsorption
2	Activated carbon (PAC/GAC*)	C	Adsorption
3	Aluminum chlorohydrates	$Al_2Cl(OH)_5$	Coagulation
4	Aluminum chloride	$AlCl_3$	Coagulation
5	Aluminum sulfate (Alum)	$Al_2(SO_4)_3$	Coagulation, phosphorus removal
6	Ammonia	NH_{3aq}	Source for preparing chloramine for disinfection
7	Ammonium sulfate	$(NH_4)_2SO_4$	Source of ammonia for preparing chloramine for disinfection
8	Calcium hydroxide	$Ca(OH)_2$	pH adjustment, softening process and corrosion control, sludge processing
9	Calcium hypochlorite	$Ca(OCl)_2$	Disinfection and oxidation
10	Calcium oxide	CaO	Coagulation aid, pH adjustment, softening, corrosion control
11	Chlorine	Cl_2	Disinfection and oxidation
12	Chloramine (mono)	NH_2Cl	Disinfection
13	Chlorine dioxide	ClO_2	Disinfection and oxidation
14	Copper sulfate	$CuSO_4$	Algae control
15	Ferric chloride	$FeCl_3$	Coagulation, sludge processing, removal of organics, phosphate and sulfide odors
16	Ferric oxide	Fe_2O_3	Adsorption

(*Continued*)

No.	Chemical	Formula	Water or wastewater treatment
17	Ferric sulfate	$Fe_2(SO_4)_3$	Coagulation, sludge processing, removal of organics, phosphates and control of sulfide odors
18	Ferrous sulfate	$FeSO_4$	Coagulation
19	Hydrochloric acid	HCl	pH adjustment
20	Hydrofluorosilicic acid	H_2SiF_6	Fluoridation
21	Hydrogen peroxide	H_2O_2	Disinfection and oxidation
22	Ozone	O_3	Disinfection and oxidation
23	Polymers (polyelectrolytes)	–	Coagulation, flocculation, sludge thickening and dewatering
24	Polyaluminum chlorides	$Al_n(OH)_mCl_{(3n-m)}$	Coagulation
25	Potassium permanganate	$KMnO_4$	Disinfection and oxidation
26	Sodium aluminate	$NaAlO_2$	Coagulation
27	Sodium bicarbonate	$NaHCO_3$	pH adjustment, softening, corrosion control
28	Sodium bisulfite	$NaHSO_3$	Dechlorination
29	Sodium carbonate	Na_2CO_3	pH adjustment, softening, corrosion control
30	Sodium fluoride	NaF	Fluoridation
31	Sodium fluorosilicate	Na_2SiF_6	Fluoridation
32	Sodium hydroxide	NaOH	pH and alkalinity adjustment, softening, corrosion control, regeneration of ion exchange resins
33	Sodium hypochlorite	NaClO	Disinfection and oxidation
34	Sodium silicate	Na_2SiO_3	Coagulation aid, flocculation aid, pH adjustment, corrosion control
35	Sodium sulfite	Na_2SO_3	Dechlorination
36	Sulfuric acid	H_2SO_4	pH adjustment
37	Sulfur dioxide	SO_2	Dechlorination
38	Zinc ortho-phosphate	$Zn_3(PO_4)_2$	Corrosion control

*PAC – Powdered Activated Carbon; GAC – Granulated Activated Carbon.

Table prepared using following reference documents.

BIBLIOGRAPHY

Alaska Department of Environmental Conservation, Division of Water (2009) Introduction to Small Water Systems – A Course for Level I Operators. http://dec.alaska.gov/water/opcert/IntroSmallCorrOnlineCourses.html (accessed 25 July 2014).

Australian Government, National Health and Medical Research Council Australian Drinking Water Guidelines 6, 2011, Version 2.0 Updated December 2013, National Water Quality Management Strategy. http://www.nhmrc.gov.au/guidelines/publications/eh52 (accessed 25 July 2014).

United States Environmental Protection Agency, Office of Water (1999) Alternative Disinfectants and Oxidants Guidance Manual. http://nepis.epa.gov/Exe/ZyPURL.cgi?Dockey=2000229L.txt (accessed 25 July 2014).

United State Environment Protection Agency, Office of Water (1999) Guidance Manual for Compliance with the Interim Enhanced Surface Water Treatment Rule: Turbidity Provisions. http://water.epa.gov/lawsregs/rulesregs/sdwa/mdbp/upload/2001_01_12_mdbp_turbidity_cover_tu.pdf (accessed 25 July 2014).

United States Environmental Protection Agency, Office of Water (1999) Wastewater Technology Fact Sheet – Ozone Disinfection. http://nepis.epa.gov/Exe/ZyPURL.cgi?Dockey=200044E8.txt (accessed 25 July 2014).

United States Environmental Protection Agency, Office of Water (2000) Wastewater Technology Fact Sheet – Dechlorination. http://nepis.epa.gov/Exe/ZyPURL.cgi?Dockey=P1001L40.txt (accessed 25 July 2014).

Water Research Foundation (2009) Supply of Critical Drinking Water and Wastewater Treatment Chemicals – A White Paper for Understanding Recent Chemical Price Increases and Shortages, Jointly Sponsored by WRF, American Water Works Association (AWWA), Water Environment Research Foundation (WERF) and U.K. Water Industry Research (UKWIR). http://www.waterrf.org/PublicReportLibrary/91264.pdf (accessed 25 July 2014).

Appendix K

Listing: Laboratory equipment and instruments

This section includes a partial list of common inventories used in water or wastewater laboratories.

Examples of common laboratory glassware, plasticware and porcelain utensils.

(1)	Beakers	(20)	Heating mantles
(2)	BOD bottles+	(21)	Micropipette
(3)	Bottle caps	(22)	Pans
(4)	Bottles	(23)	Pestle
(5)	Burettes	(24)	Pipettes
(6)	Centrifuge tubes	(25)	Porcelain casseroles
(7)	Clams and supports	(26)	Porcelain crucibles
(8)	Culture dishes	(27)	Porcelain mortar
(9)	Desiccators	(28)	Racks
(10)	Petri dishes	(29)	Sample bottles
(11)	Deionized water dispenser	(30)	Spoons, scoops, scrapers
(12)	Evaporating dishes	(31)	Spot plates
(13)	Filter papers	(32)	Stoppers
(14)	Filtering flasks	(33)	Syringes
(15)	Flasks	(34)	Trays
(16)	Funnels	(35)	Vials
(17)	Glass crucibles	(36)	Volumetric flasks
(18)	Graduated cylinders	(37)	Volumetric pipettes
(19)	Graduated pitchers	(38)	Wash Bottles

+Biochemical Oxygen Demand.

Examples of common laboratory equipment and instruments

(1)	Autoclave	(25)	Microscope
(2)	Bunsen burner	(26)	Microwave
(3)	Calorimeter	(27)	Muffle furnace
(4)	Centrifuge	(28)	Oven
(5)	Colony counter	(29)	Oxygen-reduction potential probe
(6)	Colorimeter	(30)	pH meter
(7)	Conductivity meter	(31)	Pressure gauge
(8)	Dissolved oxygen (DO) probe	(32)	Refractometer
(9)	Dry bath	(33)	Refrigerator
(10)	Analytical balance	(34)	Respirometer
(11)	Evaporator	(35)	Sampler
(12)	Filters	(36)	Sonicator
(13)	Flow meter	(37)	Spectrophotometer
(14)	Fluorometer	(38)	Surveying equipment
(15)	Freezer	(39)	TDS+ meter
(16)	Fume hood	(40)	Thermometer
(17)	Gas chromatograph	(41)	Test Kits
(18)	Homogenizer	(42)	Total organic carbon analyzer
(19)	Hot plate	(43)	Turbidimeter
(20)	Hot plate stirrer	(44)	Vacuum pump
(21)	HPLC*	(45)	Viscometer
(22)	Incubator	(46)	Vortex mixer
(23)	Magnetic stirrer	(47)	Water bath
(24)	Mass spectrometer	(48)	Weighing scale

*High Performance Liquid Chromatograph, +Total Dissolved Solids.

Examples of laboratory office, furniture, safety and miscellaneous items			
(1)	Acid resistant gloves	(19)	Lab tables, stools and chairs
(2)	Apron	(20)	Labels (reagents, materials)
(3)	Boxes	(21)	Lights
(4)	Brushes	(22)	Measuring tape
(5)	Cabinets	(23)	Paper towels
(6)	Camera	(24)	Plumbing system
(7)	Computer	(25)	Printer
(8)	Disposable gloves	(26)	Respirator
(9)	Dust Broom	(27)	Reagents
(10)	Dust Pan	(28)	Safety glasses
(11)	Eyewash	(29)	Safety showers
(12)	Fire extinguisher	(30)	Safety manual
(13)	First aid kit	(31)	Spare batteries
(14)	Forceps	(32)	Spill cleanup kit
(15)	Garbage can	(33)	Stationeries
(16)	Lab carts	(34)	Timers
(17)	Lab coats	(35)	Tongs
(18)	Lab furniture		

BIBLIOGRAPHY

Laky D., László B. and Licskó I. (2008). Effects of several water quality parameters on arsenic removal by coagulation: laboratory experiments and a pilot-scale study. *Water Practice and Technology*, **3**(3), 1–8.

Lotito V., Spinosa L., Mininni G. and Antonacci R. (1997). The rheology of sewage sludge at different steps of treatment. *Water Science and Technology*, **36**(11), 79–85.

Poutiainen H., Laitinen S., Juntunen P. and Heinonen-Tanski H. (2009). On-line microwave total solids sensoring in sewage characterization. *Water Science and Technology*, **59**(7), 1291–1297.

Pritt J. W. and Raese J. W., United States Geological Survey, National Water Quality Laboratory (1992). Quality Assurance/Quality Control Manual. http://nwql.usgs.gov/rpt.shtml?OFR-92-495 (accessed 30 July 2014).

Spanjers H., Patry G. G. and Keesman K. J. (2002). Respirometry-based on-line model parameter estimation at a full-scale WWTP. *Water Science and Technology*, **45**(4–5), 335–343.

State of Oregon Department of Environmental Quality, Laboratory and Environmental Assessment Division, USA (2013) Quality Manual. http://www.deq.state.or.us/lab/techrpts/docs/DEQ91LAB0006QMP.pdf (accessed 25 July 2014).

United States Environmental Protection Agency (2005). Manual for the Certification of Laboratories Analyzing Drinking Water – Criteria and Procedures Quality Assurance. 5th edn. http://nepis.epa.gov/Exe/ZyPURL.cgi?Dockey=30006MXP.txt (accessed 30 July 2014).

Washington State Department of Ecology (2010) Procedural Manual for the Environmental Laboratory Accreditation Program. https://fortress.wa.gov/ecy/publications/publications/1003048.pdf (accessed 25 July 2014).

Water Research Foundation (2013). Application of Filters for Evaluating Lead and Copper Concentrations in Tap Water – Assessing Risk of Lead and Copper Consumption From Drinking Water. http://www.waterrf.org/PublicReportLibrary/4415.pdf (accessed 30 July 2014).

Water Research Foundation (2013). Biological Filtration Monitoring and Control Toolbox: Guidance Manual. http://www.waterrf.org/PublicReportLibrary/4231a.pdf (accessed 30 July 2014).

Appendix L

Listing: Common construction heavy equipment

Acknowledgement

Glossary terms and their description in the present section are used with the permission from the Virginia Adult Learning Resource Center (VALRC), Virginia Commonwealth University, Virginia, USA and can be found online at http://www.valrc.org/resources/buildingbasics/index.html (accessed 24 July 2014).

No.	Equipment	General description
1	Dump truck	Equipment that hauls materials and debris away; delivers sand or other materials to a site.
2	Forklift	Equipment used for moving materials around a site and unloading trucks.
3	Grader	Equipment that makes a roadbed smooth by dragging a blade over it.
4	Compactor or roller	Equipment that smoothes the road with a heavy roller.
5	Bulldozer	Equpment that moves dirt, stones, or other materials at the site.
6	Concrete mixer	Equipment that mixes water, sand, gravel, and cement to make concrete.
7	Crane	Equipment that lifts heavy things; a tower crane lifts beams, pipes, or batches of concrete to high places.
8	Excavator or digger	Equipment that digs deep holes.
9	Front-end loader	Equipment that lifts heavy materials.
10	Backhoe loader	Equipment that uses a bucket on the back for digging and has a loader on the front for lifting.
11	Cherry picker	Equipment that lifts a worker in a bucket.
12	Outriggers or stabilizers	Part of a piece of equipment intended to maintain stability and avoid it tipping over.

Appendix M

Listing: Water and wastewater profession's key gadgets and tools

No.	Task or activity	Basic profession activities/tasks	Common reference sources
1	Project technical and financial planning	(a) Analyzing Past Planning Reports (b) Reviewing current laws and regulations (c) Determining present and future needs (d) Developing planning documents (e) Performing financial planning (f) Preparing project schedules, milestones (g) Report presentation and public review	(a) Organization's past records (b) Previous planning documents (c) Past financial planning reports (d) Client input for planning process (e) Government rules/regulations (f) Public comments data
2	Collection and distribution piping design for water and wastewater systems	(a) Reviewing existing pipes, equipment (b) Assessing pipe flow capacities (c) Performing system pressure calculations (d) Evaluating new design needs (e) Selecting pipe and equipment (f) Selecting valves, fittings and accessories (g) Reviewing material compatibility (h) Developing hydraulic profiles (i) Developing plan and profile drawings (j) Preparing project specifications	(a) Engineering mathematics books (b) Hydraulics textbooks (c) Piping handbooks (d) Pipe system manuals, standards (e) Water modeling software (f) Equipment product brochures (g) Manufacturer specifications
3	Water and wastewater pumping system designs	(a) Reviewing existing pumping system (b) Selecting new design flows, pumps (c) Developing pump hydraulics (d) Flow, pressure, head-loss calculations (e) Pump and system curves, operating point	(a) Engineering mathematics books (b) Pump station design books (c) Hydraulics textbooks (d) Field manuals and handbook (e) Instrumentation and controls book

(Continued)

No.	Task or activity	Basic profession activities/tasks	Common reference sources
		(f) Sizing wet well, inflow pipe, force-main (g) Selecting operating water elevations (h) Finalizing instrumentation and controls (i) Selecting valves, fittings and accessories	(f) Product datasheets & brochures (g) Equipment specifications (h) Manufacturer drawings
4	Water and wastewater treatment plant layout	(a) Evaluating footprint requirements (b) Comparing available technologies (c) Reviewing system layout options (d) Understanding site needs, constraints (e) Selecting treatment process (f) Finalizing equipment size, quantity (g) Estimating O & M, clearance needs (h) Site safety provisions, security (i) Developing layout plans and drawings	(a) Process engineering books (b) Layout design handbooks (c) Technology manuals (d) Equipment specifications (e) Manufacturer drawings (f) Architecture books and standards
5	Developing hydraulic profile for water and wastewater treatment plant	(a) Reviewing design flow conditions (b) Analyzing process equipment sizes (c) Selecting solids/hydraulic loadings (d) Finalizing operating water elevations (e) Deciding process pipe sizes, materials (f) Sizing weirs, decanters, splitter box (g) Performing head losses calculations (h) Assessing pipe capacities (i) Develop hydraulic profile	(a) Engineering mathematics books (b) Hydraulics textbooks (c) Product datasheets and brochures (d) Equipment specifications (e) Manufacturer drawings (f) Process operations manuals (g) Field O & M manuals
6	Treatment process equipment sizing	(a) Comparing process equipment (b) Evaluating design flow conditions (c) Estimating solids and organic loadings (d) Reviewing biological kinetics (e) Understanding chemical reactions (f) Performing sludge calculations (g) Selecting hydraulic, solid retention time (h) Selecting recycling/wasting ratios (i) Deciding electrical, mechanical needs (j) Evaluating capital and operations cost (k) Finalizing process equipment, sizes	(a) Biological systems design books (b) Environmental chemistry book (c) Electrical books and standards (d) Instruments, controls manual (e) Mechanical, plumbing manual (f) Technology Brochures (g) Sludge management manual (h) Equipment manuals (i) Equipment specifications (j) Manufacturer drawings
7	Project software modeling and system evaluations	(a) Comparing process software (b) Selecting modeling tools (c) Performing simulation and modeling (d) Developing results and recommendations	(a) Software design manuals (b) Simulation modeling books

No.	Task or activity	Basic profession activities/tasks	Common reference sources
8	Process aeration system designs	(a) Performing air pipe sizing calculations (b) Determining air head-loss/pressures (c) Evaluating oxygen uptake rates (d) Selecting distribution pipe configurations (e) Selecting blower sizes and layouts (f) Reviewing electrical, controls needs (g) Finalizing drawings, specifications	(a) Aeration system design books (b) Air pipe materials catalog (c) Equipment specifications (d) Manufacturer drawings (e) Electrical books and manuals (f) Instrumentation and controls book
9	Water and wastewater systems instrumentation and controls designs	(a) Reviewing process control requirements (b) Evaluating system automation needs (c) Identifying, listing control points (d) Reviewing alarms, lights, sirens, and so on (e) Selecting process control equipment (f) Selecting process instrumentation	(a) Electrical engineering books (b) Instrumentation books (c) Process equipment manuals (d) Equipment specifications (e) Manufacturer drawings
10	Water and wastewater treatment plant buildings	(a) Reviewing building layout, architecture (b) Lab, lunch, meeting rooms (c) Building interior and exterior finishes (d) Room clearances, spacing (e) Electrical and mechanical needs (f) Lighting, heating, cooling, ventilation (g) Utilities, furniture, amenities, and so on (h) Laboratory set-up and layout (i) Safety and emergency arrangements (j) Restrooms, building access needs (k) Service roads, parking areas, and so on	(a) Architecture design books (b) Building manuals (c) Equipment specifications (d) Manufacturer drawings (e) Laboratory design manuals (f) Electrical engineering books (g) Instrumentation books (h) Mechanical and plumbing manuals (i) Safety manuals (j) Traffic regulations
11	Water and wastewater system operations and maintenance	(a) Performing routine operations (b) Preparing daily, monthly, annual logs (c) Developing operations, finance reports (d) Obtaining, maintaining permits (e) Recording system maintenance events (f) Managing chemicals and other inventories (g) Reporting system violations (h) Updating emergency response needs	(a) Standard operating procedures (b) O & M manuals (c) Operations log sheets (d) Safety and emergency procedures (e) Government regulations and laws
12	Water and wastewater laboratory operations	(a) Collecting, transporting, storing samples (b) Preparing chemical reagents, solutions	(a) Laboratory operations manuals (b) Standard operating procedures (c) Standard methods handbooks (d) Laboratory equipment manuals

(*Continued*)

No.	Task or activity	Basic profession activities/tasks	Common reference sources
		(c) Operating lab equipment and instruments (d) Performing laboratory analysis (e) Managing laboratory inventory (f) Ensuring lab safety (g) Maintaining personnel safety gears (h) Maintaining chemical data sheets (i) Properly disposing hazardous chemicals (j) Preparing analytical reports (k) Presenting data at meetings, conferences	(e) Manufacturer drawings (f) Equipment specifications (g) Safety guidance manuals (h) Chemical suppliers contacts
13	Water and wastewater research work	(a) Laboratory scale studies (b) Bench-scale and pilot-scale research and experiments (c) Reactor, lab equipment operations (d) Sampling and analysis (e) Report preparations (f) Research publications (g) Presentations and seminars (h) Technical conferences	(a) Literature reviews (b) Research reports and papers (c) Laboratory operations manuals (d) Equipment manuals (e) Standard methods handbook (f) Process technology books (g) Safety guidance manuals
14	Project cost estimations	(a) Obtaining proposals from manufacturers (b) Preparing life cycle cost analysis (c) Selecting appropriate technology (d) Estimating engineering, services fees (e) Identifying equipment sizes, costs (f) Pipes, fittings inventory costs (g) Electrical, instrumentation costs (h) Structural, mechanical, plumbing costs (i) Construction and installation costs (j) Process start-up costs (k) Identifying permit needs, related costs (l) Building requirements and costs (m) Reviewing project contingencies needs (n) Developing projects cost spreadsheets	(a) Cost estimation handbooks (b) Manufacturer Quotes (c) Past Projects Cost Estimates

BIBLIOGRAPHY

Alaska Department of Environmental Conservation, Division of Water (2013). Introduction to Small Water Systems Course Manual. http://dec.alaska.gov/water/opcert/IntroSmallCorrOnlineCourses.html (accessed 28 July 2014).

Ashley K., Mavinic D. and Koch F. (2009). International Conference on Nutrient Recovery from Wastewater Streams. International Water Association Publishing, UK.

Butler, D. and Memon Fayyaz Ali (2006). Water Demand Management Publication. International Water Association Publishing, UK.

Commonwealth of Massachusetts, Division of Capital Asset Management (2006). Consultants Estimating Manual. http://www.mass.gov/anf/docs/dcam/dlforms/cem-feb06.pdf (accessed 30 July 2014).

Dong X., Zeng S. and Chen J. (2012). A spatial multi-objective optimization model for sustainable urban wastewater system layout planning. *Water Science and Technology*, **66**(2), 267–274.

Ellis B. and Bertrand-Kranjewski J. (2010). Assessing Infiltration and Exfiltration on the Performance of Urban Sewer Systems (APUSS). International Water Association Publishing, UK.

Fleming Training Center (2011). Division of Water Resources, Tennessee Department of Environment and Conservation, USA Wastewater Treatment Mathematical Formulas. 2nd edn. http://www.tn.gov/environment/water/fleming_math-books.shtml (accessed 24 July 2014).

Hao X., Novotny V. and Nelson V. (2010). Water Infrastructure for Sustainable Communities – China and the World, Cities of the Future Series. International Water Association Publishing, UK.

Henze M., Loosdrecht M., Ekama G. and Brdjanovic D. (2008). Biological Wastewater Treatment Principals, Modelling and Design. International Water Association Publishing, UK.

Hughes J., Eskaf S. and Thorsten R. (2006). Multi-level financial analysis of residential water and wastewater rates and rate-setting practices. *Water Utility Management International*, **1**(4), 18–22.

Kentucky Division of Compliance Assistance, Energy and Environment Cabinet, Department of Environmental Protection, USA Wastewater Treatment Conversion Factors and Formulas. http://dca.ky.gov/OPCert/certification/Documents/WWConversionTable[1].pdf (accessed 24 July 2014).

Lamont H. and Young K. (2007). Guide to Developing a Municipal Wastewater Project – for Small, Rural Communities in New York State, Published by Lamont Engineers, New York, USA. (www.lamontengineers.com/documents/GuidanceManual.pdf (accessed 30 July 2014).

Larsen T., Udert K. and Lienert J. (2013). Source Separation and Decentralization for Wastewater Management. IWA Publishing, UK.

Liang X. and van Dijk M. P. (2010). Financial and economic feasibility of decentralized wastewater reuse systems in Beijing. *Water Science and Technology*, **61**(8), 1965–1973.

Lombardo P. (2004) Cluster Wastewater Systems Planning Handbook. Project No. WU-HT-01-45, Prepared for the National Decentralized Water Resources Capacity Development Project, Washington University, St. Louis, MO, by Lombardo Associates, Inc., MA, USA. (www.ndwrcdp.org/documents/WU-HT-01-45/WUHT0145_web1.pdf (accessed 30 July 2014).

Makinia J. (2010). Mathematical Modelling and Computer Simulations of Activated Sludge Systems. IWA Publishing, UK.

Ohio Environmental Protection Agency (2005). Division of Drinking and Ground Waters, USA Applied Wastewater Math Formula Sheet and Conversion Factors. http://www.epa.ohio.gov/portals/28/documents/opcert/WWformulas.pdf (accessed 24 July 2014).

Pollard S. J. T. (2008). Risk Management for Water and Wastewater Utilities. International Water Association Publishing, UK.

State of Connecticut, Department of Public Health Drinking Water Section (2008). Liquid Chemical Feed System Design and Installation Guidelines. http://www.ct.gov/dph/lib/dph/drinking_water/pdf/Liquid_Chemical_Feed_System_Design_and_Construction_Guidelines.pdf (accessed 28 July 2014).

State of Oregon Department of Environmental Quality (DEQ), Laboratory and Environmental Assessment Division, USA (2013) Quality Manual. http://www.deq.state.or.us/lab/techrpts/docs/DEQ91LAB0006QMP.pdf (accessed 25 July 2014).

United States Environmental Protection Agency, Office of Water (1999). Guidance Manual for Compliance with the Interim Enhanced Surface Water Treatment Rule: Turbidity Provisions. http://nepis.epa.gov/Exe/ZyPURL.cgi?Dockey=200021OB.txt (accessed 28 July 2014).

United States Environmental Protection Agency (2000). Wastewater Technology Fact Sheet: Pipe Construction and Materials. http://nepis.epa.gov/Exe/ZyPURL.cgi?Dockey=P1000W83.txt (accessed 30 July 2014).

United States Environmental Protection Agency (2000). Wastewater Technology Fact Sheet: Sewers, Force Main. http://nepis.epa.gov/Exe/ZyPURL.cgi?Dockey=P10099PU.txt (accessed 30 July 2014).

United States Environmental Protection Agency, Office of Water (2000). Collection Systems Technology Fact Sheet: Sewers, Lift Station. http://nepis.epa.gov/Exe/ZyPURL.cgi?Dockey=901U0X00.txt (accessed 28 July 2014).

United States Environmental Protection Agency (2002). Collection Systems Technology Fact Sheet: Sewers, Conventional Gravity. http://nepis.epa.gov/Exe/ZyPURL.cgi?Dockey=P10053D9.txt (accessed 30 July 2014).

United States Environmental Protection Agency (2002). Wastewater Technology Fact Sheet: Sewers, Pressure. http://nepis.epa.gov/Exe/ZyPURL.cgi?Dockey=P10099Q2.txt (accessed 30 July 2014).

United States Environmental Protection Agency (2005). Handbook for Managing Onsite and Clustered (Decentralized) Wastewater Treatment Systems. http://nepis.epa.gov/Exe/ZyPURL.cgi?Dockey=20017K2G.txt (accessed 30 July 2014).

United States Environmental Protection Agency (2005). Manual for the Certification of Laboratories Analyzing Drinking Water – Criteria and Procedures Quality Assurance. 5th edn http://nepis.epa.gov/Exe/ZyPURL.cgi?Dockey=30006MXP.txt (accessed 30 July 2014)

United States Environmental Protection Agency (2006). Water System Operator Roles and Responsibilities: A Best Practices Guide. http://nepis.epa.gov/Exe/ZyPURL.cgi?Dockey=2000ZZBC.txt (accessed 30 July 2014).

United States Environmental Protection Agency, Office of Water (2006) Emerging Technologies for Biosolids Management. http://nepis.epa.gov/Exe/ZyPURL.cgi?Dockey=P1006DGM.txt (accessed 28 July 2014).

United States Environmental Protection Agency, Office of Water (2008). Onsite Wastewater Treatment Systems Manual. http://nepis.epa.gov/Exe/ZyPURL.cgi?Dockey=30004GXI.txt (accessed 28 July 2014).

van Haandel A. C. and van der Lubbe J. G. M. (2012). Handbook of Biological Wastewater Treatment – Design and Optimization of Activated Sludge System Book. 2nd edn, IWA Publishing, UK.

Washington State Department of Transportation (WSDOT), Environmental and Engineering Programs, Strategic Analysis and Estimation Office (2009). Cost Estimating Manual for WSDOT Projects, Administrative Manual. http://www.wsdot.wa.gov/publications/manuals/fulltext/m3034/estimatingguidelines.pdf (accessed 30 July 2014).

Washington State Department of Ecology (2010). Procedural Manual for the Environmental Laboratory Accreditation Program. https://fortress.wa.gov/ecy/publications/publications/1003048.pdf (accessed 25 July 2014).

Appendix N

Water project cost estimation sheets

Table N.1 Sample project cost estimation sheet for water treatment systems.

No.	Cost item	Quantity	Units	Unit cost	Total cost
1	Surveying and planning costs	———	EA	$ ———	$ ———
2	Engineering, architectural, consulting fees	———	EA	$ ———	$ ———
3	Permits and licenses	———	EA	$ ———	$ ———
4	Bonds and insurances	———	EA	$ ———	$ ———
5	Mobilization and demobilization	———	EA	$ ———	$ ———
6	Traffic management costs	———	EA	$ ———	$ ———
7	Erosion and sedimentation control measures	———	EA	$ ———	$ ———
8	Water system (tank, equipment and installation)	———	EA	$ ———	$ ———
	• Water reservoir and intake structures	———	EA	$ ———	$ ———
	• Influent pumping station	———	EA	$ ———	$ ———
	• Rapid mix and flocculation systems	———	EA	$ ———	$ ———
	• Sedimentation and settling processes	———	EA	$ ———	$ ———
	• Filtration systems	———	EA	$ ———	$ ———
	• Ion exchange and adsorption units	———	EA	$ ———	$ ———
	• Disinfection systems	———	EA	$ ———	$ ———
	• Clear well storage system	———	EA	$ ———	$ ———
	• Chemical storage and feed systems	———	EA	$ ———	$ ———
	• Aeration blower packages	———	EA	$ ———	$ ———
	• Various water conveyance pumps	———	EA	$ ———	$ ———
	• Walkways, grating, handrails, access needs	———	EA	$ ———	$ ———

(*Continued*)

Table N.1 Sample project cost estimation sheet for water treatment systems (*Continued*).

No.	Cost item	Quantity	Units	Unit cost	Total cost
9	Yard piping (equipment, installation)	———	EA	$ ———	$ ———
	• Process water gravity piping, force-mains	———	Length	$ ———	$ ———
	• Utility water piping	———	Length	$ ———	$ ———
	• Air header and distribution piping	———	Length	$ ———	$ ———
	• Isolation and control valves	———	EA	$ ———	$ ———
	• Fittings (wye, bends, tees, reducers, etc.)	———	EA	$ ———	$ ———
10	Building construction costs	———	Area	$ ———	$ ———
	• Administrative building	———	Area	$ ———	$ ———
	• Mechanical building	———	Area	$ ———	$ ———
	• Process control building	———	Area	$ ———	$ ———
	• Laboratory building	———	Area	$ ———	$ ———
11	HVAC, electrical, instrumentation and controls costs	———	EA	$ ———	$ ———
12	Mechanical and plumbing costs	———	EA	$ ———	$ ———
13	Safety, fire protection, emergency response, security	———	EA	$ ———	$ ———
14	Earthworks, clearing and demolition costs	———	EA	$ ———	$ ———
15	Construction management and start-up costs	———	EA	$ ———	$ ———
16	Project miscellaneous items costs	———	EA	$ ———	$ ———
17	Contingencies	———	EA	$ ———	$ ———
Total project costs					$ ———

EA – Each Unit or Item and HVAC – Heating Ventilating and Air Conditioning.

Table N.2 Sample project cost estimation sheet for water distribution systems.

No.	Cost item	Quantity	Units	Unit cost	Total cost
1	Surveying and planning costs	———	EA	$ ———	$ ———
2	Engineering, architectural, consulting fees	———	EA	$ ———	$ ———
3	Permits and licenses	———	EA	$ ———	$ ———
4	Bonds and insurances	———	EA	$ ———	$ ———
5	Mobilization and demobilization	———	EA	$ ———	$ ———
6	Traffic management costs	———	EA	$ ———	$ ———
7	Erosion and sedimentation control measures	———	EA	$ ———	$ ———

Table N.2 Sample project cost estimation sheet for water distribution systems (*Continued*).

No.	Cost Item	Quantity	Units	Unit cost	Total cost
8	Water distribution (tank, equipment and installation)	——	EA	$ ——	$ ——
	• Water pressure piping network	——	Length	$ ——	$ ——
	• Water storage reservoirs	——	EA	$ ——	$ ——
	• At-grade water storage tanks	——	EA	$ ——	$ ——
	• Elevated water storage tanks or towers	——	EA	$ ——	$ ——
	• Underground storage tanks, cisterns	——	EA	$ ——	$ ——
	• Pumping stations	——	EA	$ ——	$ ——
	• Pressure vessels	——	EA	$ ——	$ ——
	• Fire hydrants	——	EA	$ ——	$ ——
	• Flow meters	——	EA	$ ——	$ ——
	• Isolation and control valves	——	EA	$ ——	$ ——
	• Fittings (wye, bends, tees, reducers, etc.)	——	EA	$ ——	$ ——
	• Cleanouts, flushing systems	——	EA	$ ——	$ ——
	• Distribution process controls	——	EA	$ ——	$ ——
9	Electrical and instrumentation costs	——	EA	$ ——	$ ——
10	Mechanical and plumbing costs	——	EA	$ ——	$ ——
11	Safety, fire protection, emergency response, security	——	EA	$ ——	$ ——
12	Earthworks, clearing and demolition costs	——	EA	$ ——	$ ——
13	Construction management and start-up costs	——	EA	$ ——	$ ——
14	Project miscellaneous items costs	——	EA	$ ——	$ ——
15	Contingencies	——	EA	$ ——	$ ——
Total project costs					$ ——

EA – Each Unit or Item.

BIBLIOGRAPHY

Alaska Department of Environmental Conservation, Division of Water (2013). Introduction to Small Water Systems Course Manual. http://dec.alaska.gov/water/opcert/IntroSmallCorrOnlineCourses.html (accessed 28 July 2014).

Commonwealth of Massachusetts, Division of Capital Asset Management (2006). Consultants Estimating Manual. http://www.mass.gov/anf/docs/dcam/dlforms/cem-feb06.pdf (accessed 30 July 2014).

Lamont H. and Young K. (2007). Guide to Developing a Municipal Wastewater Project – for Small, Rural Communities in New York State, Published by Lamont Engineers, New York, USA. www.lamontengineers.com/documents/GuidanceManual.pdf (accessed 30 July 2014).

Safe Drinking Water Foundation, Canada (2009). Conventional Water Treatment: Coagulation and Filtration Factsheet. www.safewater.org/resources/fact-sheets.html (accessed 30 July 2014).

United States Environmental Protection Agency (2004). Safe Drinking Water Act Factsheet: Drinking Water Treatment. http://nepis.epa.gov/Exe/ZyPURL.cgi?Dockey=3000660S.txt (accessed 30 July 2014).

United States Environmental Protection Agency (2006). Water System Operator Roles and Responsibilities: A Best Practices Guide. http://nepis.epa.gov/Exe/ZyPURL.cgi?Dockey=2000ZZBC.txt (accessed 30 July 2014).

United States Environmental Protection Agency, Office of Water (2009). Drinking Water Treatment Guidance Document. http://nepis.epa.gov/Exe/ZyPURL.cgi?Dockey=3000660S.txt (accessed 28 July 2014).

United States Environmental Protection Agency Water: Private Wells Section. http://water.epa.gov/drink/info/well/basicinformation.cfm (accessed July 2014).

United States Fire Administration (2008). Water Supply Systems and Evaluation Methods, Volume I: Water Supply System Concepts. www.usfa.fema.gov/fireservice/prevention_education/strategies/water_supply (accessed 28 July 2014).

Washington State Department of Transportation, Environmental and Engineering Programs, Strategic Analysis and Estimation Office (2009). Cost Estimating Manual for WSDOT Projects, Administrative Manual. http://www.wsdot.wa.gov/publications/manuals/fulltext/m3034/estimatingguidelines.pdf (accessed 30 July 2014).

Appendix O

Wastewater project cost estimation sheets

Table O.1 Sample project cost sheet for sewer collection and conveyance systems.

No.	Cost item	Quantity	Units	Unit cost	Total cost
1	Surveying and planning costs	———	EA	$ ———	$ ———
2	Engineering, architectural, consulting fees	———	EA	$ ———	$ ———
3	Permits and licenses	———	EA	$ ———	$ ———
4	Bonds and insurances	———	EA	$ ———	$ ———
5	Mobilization and demobilization	———	EA	$ ———	$ ———
6	Traffic management costs	———	EA	$ ———	$ ———
7	Erosion and sedimentation control measures	———	EA	$ ———	$ ———
8	Sewer collection system (equipment and installation)	———	EA	$ ———	$ ———
	• Gravity sewer piping	———	Length	$ ———	$ ———
	• Pressure sewer piping	———	Length	$ ———	$ ———
	• Vacuum sewer piping	———	Length	$ ———	$ ———
	• Grinder pump units	———	EA	$ ———	$ ———
	• Vacuum station	———	EA	$ ———	$ ———
	• Manholes, vaults, accessories	———	EA	$ ———	$ ———
	• Isolation, control and air release valves	———	EA	$ ———	$ ———
	• Fittings (wye, bends, tees, reducers, etc.)	———	EA	$ ———	$ ———
	• Cleanouts, flushing systems	———	EA	$ ———	$ ———

(*Continued*)

Table O.1 Sample project cost sheet for sewer collection and conveyance systems (*Continued*).

No.	Cost item	Quantity	Units	Unit cost	Total cost
9	Sewer pump stations (equipment and installation)	———	EA	$ ———	$ ———
	• Wet well and dry well basins	———	EA	$ ———	$ ———
	• Pump assemblies, base, guide rails, hoist	———	EA	$ ———	$ ———
	• Floats, level sensors, lights, alarms	———	EA	$ ———	$ ———
	• Pump controls, panels, flow meter	———	EA	$ ———	$ ———
	• Valve vault, valves, fittings	———	EA	$ ———	$ ———
	• Walkway, grating, access openings	———	EA	$ ———	$ ———
10	Septic systems (equipment and installation)	———	EA	$ ———	$ ———
	• Septic tank, filter, baffle, access opening	———	EA	$ ———	$ ———
	• Absorption area, distribution piping	———	EA	$ ———	$ ———
	• Sludge hauling	———	EA	$ ———	$ ———
11	Electrical, instrumentation and controls costs	———	EA	$ ———	$ ———
12	Mechanical and plumbing costs	———	EA	$ ———	$ ———
13	Safety, fire protection, emergency response, security	———	EA	$ ———	$ ———
14	Earthworks, clearing and demolition costs	———	EA	$ ———	$ ———
15	Construction management and start-up costs	———	EA	$ ———	$ ———
16	Project miscellaneous items costs	———	EA	$ ———	$ ———
17	Contingencies	———	EA	$ ———	$ ———
Total project costs					$ ———

EA – Each Unit or Item.

Table O.2 Sample project cost sheet for wastewater treatment systems.

No.	Cost item	Quantity	Units	Unit cost	Total cost
1	Surveying and planning costs	———	EA	$ ———	$ ———
2	Engineering, architectural, consulting fees	———	EA	$ ———	$ ———
3	Permit and licenses	———	EA	$ ———	$ ———
4	Bonds and insurances	———	EA	$ ———	$ ———
5	Mobilization and demobilization	———	EA	$ ———	$ ———
6	Traffic management costs	———	EA	$ ———	$ ———
7	Erosion and sedimentation control measures	———	EA	$ ———	$ ———

(*Continued*)

Table O.2 Sample project cost sheet for wastewater treatment systems (*Continued*).

No.	Cost item	Quantity	Units	Unit cost	Total cost
8	Wastewater treatment (tank, equipment, installation)	———	EA	$ ———	$ ———
	• Influent pump station, vault, flow meter	———	EA	$ ———	$ ———
	• Screening units (bar screen, fine screen)	———	EA	$ ———	$ ———
	• Grinder systems	———	EA	$ ———	$ ———
	• Grit removal and clarification units	———	EA	$ ———	$ ———
	• Biological treatment systems	———	EA	$ ———	$ ———
	• Disinfection systems	———	EA	$ ———	$ ———
	• Sludge thickening and stabilization systems	———	EA	$ ———	$ ———
	• Sludge dewatering and drying units	———	EA	$ ———	$ ———
	• Process pumps (wastewater, sludge, water)	———	EA	$ ———	$ ———
	• Chemical storage and feed systems	———	EA	$ ———	$ ———
	• Aeration blowers, mixers	———	EA	$ ———	$ ———
	• Walkways, grating, access areas	———	EA	$ ———	$ ———
9	Wastewater yard piping (including installation)	———	EA	$ ———	$ ———
	• Wastewater gravity piping, force-mains	———	Length	$ ———	$ ———
	• Utility and plant water piping	———	Length	$ ———	$ ———
	• Air header and distribution piping	———	Length	$ ———	$ ———
	• Manholes, access openings	———	EA	$ ———	$ ———
	• Isolation and control valves	———	EA	$ ———	$ ———
	• Fittings (wye, bends, tees, reducers, etc.)	———	EA	$ ———	$ ———
	• Cleanouts, flushing systems	———	EA	$ ———	$ ———
10	Building construction costs	———	Area	$ ———	$ ———
	• Administrative building	———	Area	$ ———	$ ———
	• Headworks building	———	Area	$ ———	$ ———
	• Mechanical building	———	Area	$ ———	$ ———
	• Process control building	———	Area	$ ———	$ ———
	• Laboratory building	———	Area	$ ———	$ ———
11	HVAC, electrical, instrumentation and controls costs	———	EA	$ ———	$ ———
12	Mechanical and plumbing costs	———	EA	$ ———	$ ———
13	Safety, fire protection, emergency response, security	———	EA	$ ———	$ ———
14	Earthworks, clearing and demolition costs	———	EA	$ ———	$ ———
15	Construction management and start-up costs	———	EA	$ ———	$ ———
16	Project miscellaneous items costs	———	EA	$ ———	$ ———
17	Contingencies	———	EA	$ ———	$ ———
Total project costs					$ ———

EA – Each Unit or Item and HVAC – Heating Ventilating and Air Conditioning.

BIBLIOGRAPHY

Commonwealth of Massachusetts, Division of Capital Asset Management (2006). Consultants Estimating Manual. http://www.mass.gov/anf/docs/dcam/dlforms/cem-feb06.pdf (accessed 30 July 2014).

Lamont H. and Young K. (2007). Guide to Developing a Municipal Wastewater Project – for Small, Rural Communities in New York State, Published by Lamont Engineers, New York, USA. www.lamontengineers.com/documents/GuidanceManual.pdf (accessed 30 July 2014).

Libabher M. and Orozco-Jaramillo A. (2012). Sustainable Treatment and Reuse of Municipal Wastewater. IWA Publishing, UK.

Lombardo P. (2004). Cluster Wastewater Systems Planning Handbook. Project No. WU-HT-01-45, Prepared for the National Decentralized Water Resources Capacity Development Project, Washington University, St. Louis, MO, by Lombardo Associates, Inc., MA, USA. www.ndwrcdp.org/documents/WU-HT-01-45/WUHT0145_web1.pdf (accessed 30 July 2014).

United States Environmental Protection Agency, Office of Water (2000). Collection Systems Technology Fact Sheet: Sewers, Lift Station. http://nepis.epa.gov/Exe/ZyPURL.cgi?Dockey=901U0X00.txt (accessed 28 July 2014).

United States Environmental Protection Agency (2000). Wastewater Technology Fact Sheet: Pipe Construction and Materials. http://nepis.epa.gov/Exe/ZyPURL.cgi?Dockey=P1000W83.txt (accessed 30 July 2014).

United States Environmental Protection Agency (2000). Wastewater Technology Fact Sheet: Sewers, Force Main. http://nepis.epa.gov/Exe/ZyPURL.cgi?Dockey=P10099PU.txt (accessed 30 July 2014).

United States Environmental Protection Agency (2002). Collection Systems Technology Fact Sheet: Sewers, Conventional Gravity. http://nepis.epa.gov/Exe/ZyPURL.cgi?Dockey=P10053D9.txt (accessed 30 July 2014).

United States Environmental Protection Agency (2002). Wastewater Technology Fact Sheet: Sewers, Pressure. http://nepis.epa.gov/Exe/ZyPURL.cgi?Dockey=P10099Q2.txt (accessed 30 July 2014).

United States Environmental Protection Agency (2005). Handbook for Managing Onsite and Clustered (Decentralized) Wastewater Treatment Systems. http://nepis.epa.gov/Exe/ZyPURL.cgi?Dockey=20017K2G.txt (accessed 30 July 2014).

United States Environmental Protection Agency, Office of Water (2008) Onsite Wastewater Treatment Systems Manual. http://nepis.epa.gov/Exe/ZyPURL.cgi?Dockey=30004GXI.txt (accessed 28 July 2014).

van Haandel A. C. and van der Lubbe J. G. M. (2012). Handbook of Biological Wastewater Treatment – Design and Optimization of Activated Sludge System. 2nd edn, IWA Publishing, UK.

Washington State Department of Transportation, Environmental and Engineering Programs, Strategic Analysis and Estimation Office (2009). Cost Estimating Manual for WSDOT Projects, Administrative Manual. http://www.wsdot.wa.gov/publications/manuals/fulltext/m3034/estimatingguidelines.pdf (accessed 30 July 2014).

Appendix P

Laboratory project cost estimation sheet

Table P.1 Sample project cost sheet for water or wastewater laboratory.

No.	Cost item	Quantity	Units	Unit cost	Total cost
1	Surveying and planning costs	———	EA	$ ———	$ ———
2	Engineering, architectural, consulting fees	———	EA	$ ———	$ ———
3	Permits and licenses	———	EA	$ ———	$ ———
4	Bonds and insurances	———	EA	$ ———	$ ———
5	Traffic management costs	———	EA	$ ———	$ ———
6	Erosion and sedimentation control measures	———	EA	$ ———	$ ———
7	Mobilization and demobilization	———	EA	$ ———	$ ———
8	Laboratory building construction cost	———	EA	$ ———	$ ———
	• Building construction	———	EA	$ ———	$ ———
	• Lab interior construction	———	EA	$ ———	$ ———
	• Lab interior and exterior finishes	———	EA	$ ———	$ ———
	• Lunch, office and meeting rooms	———	EA	$ ———	$ ———
	• Plumbing, piping and mechanical systems	———	EA	$ ———	$ ———
	• HVAC, electrical system, lighting	———	EA	$ ———	$ ———
	• Fire protection system	———	EA	$ ———	$ ———
9	Lab furniture and office amenities	———	EA	$ ———	$ ———
	• Lab shelves and cabinets	———	EA	$ ———	$ ———
	• Laboratory operation work stations	———	EA	$ ———	$ ———
	• Benches, chairs and stools	———	EA	$ ———	$ ———
	• Office working desk	———	EA	$ ———	$ ———
	• Office equipment, file cabinets, and so on	———	EA	$ ———	$ ———
	• Computers and lab software	———	EA	$ ———	$ ———
	• Books, manuals, standards, and so on	———	EA	$ ———	$ ———
	• Water sink, trash cans, shredders	———	EA	$ ———	$ ———

(*Continued*)

Table P.1 Sample project cost sheet for water or wastewater laboratory (*Continued*).

No.	Cost item	Quantity	Units	Unit cost	Total cost
10	Laboratory safety equipment installation	——	EA	$ ——	$ ——
	• Fuming hood	——	EA	$ ——	$ ——
	• Safety showers, eyewash	——	EA	$ ——	$ ——
	• Apron, gloves, safety glasses, and so on	——	EA	$ ——	$ ——
	• Hazardous material disposal systems	——	EA	$ ——	$ ——
11	Laboratory equipment, instruments and chemicals	——	EA	$ ——	$ ——
	• Laboratory equipment	——	EA	$ ——	$ ——
	• Analytical instruments	——	EA	$ ——	$ ——
	• Lab glassware, plasticware, and so on	——	EA	$ ——	$ ——
	• Chemicals, biochemicals, reagents	——	EA	$ ——	$ ——
	• Other lab supplies and storage provisions	——	EA	$ ——	$ ——
12	Fire protection, emergency response, security	——	EA	$ ——	$ ——
13	Earthworks, clearing and demolition costs	——	EA	$ ——	$ ——
14	Construction management and start-up costs	——	EA	$ ——	$ ——
15	Project miscellaneous costs	——	EA	$ ——	$ ——
16	Contingencies	——	EA	$ ——	$ ——
Total project costs					$ ——

EA – Each Unit or Item.

BIBLIOGRAPHY

Commonwealth of Massachusetts, Division of Capital Asset Management (2006). Consultants Estimating Manual. http://www.mass.gov/anf/docs/dcam/dlforms/cem-feb06.pdf (accessed 30 July 2014).

Lamont H. and Young K. (2007). Guide to Developing a Municipal Wastewater Project – for Small, Rural Communities in New York State, Published by Lamont Engineers, New York, USA. www.lamontengineers.com/documents/GuidanceManual.pdf (accessed 30 July 2014).

Pritt J. W. and Raese J. W. and United States Geological Survey (2012). Quality Assurance/Quality Control Manual. http://nwql.usgs.gov/rpt.shtml?OFR-92-495 (accessed 30 July 2014).

State of Oregon Department of Environmental Quality, Laboratory and Environmental Assessment Division, USA (2013). Quality Manual. http://www.deq.state.or.us/lab/techrpts/docs/DEQ91LAB0006QMP.pdf (accessed 25 July 2014).

United States Environmental Protection Agency (2005). Manual for the Certification of Laboratories Analyzing Drinking Water – Criteria and Procedures Quality Assurance. 5th edn, http://nepis.epa.gov/Exe/ZyPURL.cgi?Dockey=30006MXP.txt (accessed 30 July 2014).

Washington State Department of Ecology (2010). Procedural Manual for the Environmental Laboratory Accreditation Program. https://fortress.wa.gov/ecy/publications/publications/1003048.pdf (accessed 25 July 2014).

Washington State Department of Transportation, Environmental and Engineering Programs, Strategic Analysis and Estimation Office (2009). Cost Estimating Manual for WSDOT Projects, Administrative Manual. http://www.wsdot.wa.gov/publications/manuals/fulltext/m3034/estimatingguidelines.pdf (accessed 30 July 2014).

Index

IWA Publishing's authorised EU representative for General Product Safety Regulations is Diane D'Arras, 15 rue Duret, 75116 Paris, France, e-mail: safety@iwap.co.uk.

Printed and bound by CPI Group (UK) Ltd, Croydon, CR0 4YY
12/05/2026
02108866-0010